A new environmental planning model.

<u>Technology and Innovation - Research and Development,</u>

<u>A challenge for millennials generations against global warming.</u>

ALEXIS JOSE LOPEZ DELGADO

A new environmental planning model

Copyright © 2020 Alexis Jose Lopez Delgado

All rights reserved.

ISBN: 9798571628679

DEDICATION

To God the Father, Son, and Holy Spirit.

CONTENT

Thanks I

1 Chapter I

    A Brief History of Mining .... Page No.4

    The oldest mine.

    Grimes's Graves of the Neolithic.

    The methods of metalworking in ancient Egypt.

    Mining at the height of the Roman Empire.

    Urban development in ancient Greece.

    Mining, the advance of the Middle Ages.

2 Chapter II

    A Brief History of the Oil Industry .... No. 22

    Oil Formation.

    Pre-Columbian Oil.

    Asia and Europe with the first hydrocarbon deposits.

3 Chapter III

    Exploration and Exploitation .... Page No. 31

    Scanning tools and techniques.

4 Chapter IV

    The role of exploration and exploitation research in the mining and oil industries. .... No. 45

4.1 Mining .... P. 56

Development of research and application of innovation in mining.

4.2 Oil ....

No. 51

Innovation and technology need for the oil and gas industry.

Great challenges of the oil and gas industry.

Technology in upstream, midstream and downstream.

Greenhouse gas mitigation.

5 Chapter V

Scientific research for the impulse of human development...

No. 61

Latin American energy savings.

6 Chapter VI

A new planning model .... No. 68

Strategic environmental planning.

T&I -I&D and environment.

Latin America and the Caribbean and environmental technologies.

Pending points to integrate and include.

7 Conclusion   No. 82

8 References   No. 84

A new environmental planning model

THANKS

To all the institutions that gave me the academic training and wisdom to transmit knowledge.
To all my colleagues and friends who have somehow or other collaborated in the search for innovative information and ideas.
To the companies that I have belonged to by providing me with the opportunity and confidence to be in them, managing their economic and human resources.
To my family for donating their time, thus allowing me to write these lines.
Thank you all.

INTRODUCTION.

The need to apply scientific and technological changes in a *green sense with* innovation and application of their results in the world, from thinking, design, production and that consumers themselves generate solutions, products and benefits without consequences, is the obligation of our generations. Technological demands, in a world where it becomes increasingly dependent on the energy resources present in the global supply, have given some solutions, and on the sidewalk in front; creativity and the desire to offer new models of improvement, technology conservation and knowledge transfer have indisputably advanced. However, it is necessary to reduce energy consumption and save in industrial maintenance, in the oil and mining area, adjugating the conservation and improvement of the recovery of non-renewable natural resources, optimizing their consumption. In recent decades, several concepts of poor understanding have been raised, ultimately reflecting the interest in sustaining life on our planet. These concepts are: sustainable development and eco-development. But what are these sustainable technologies that have given or propose sociotechnical solutions, in a complex society where changes are constant. There are automations, accelerated demands, high demands in quality in the shortest possible time, and technical production is focused on substantial and commercial savings in imbalance with the well-being of our societies and our planet. The need to innovate in maintenance technologies as a model of endogenous, equitable, sustainable, ecological, solidarity and distributional development for majority is evident.

1 BRIEF HISTORY OF MINING

The mining industry is one of the oldest in humanity and has long history, since the past times, with our first ancestors, The inhabitants of the prehistoric era were, without a doubt, absolute miners man has used various minerals to manufacture work utensils, weapons for hunting or defense, even for its aesthetics, anthropological studies hold us that many types of minerals and precious stones, including gold, where the **latter was one of the first metals to hook the anachronistic man,** not only for his striking beauty, but also for his condition of malleability to facilitate the making of ornaments that highlighted the social or hierarchical condition of his possessor.

Over time, mineral extraction became an important industry that has created physical-chemical techniques, studies and analysis to improve the exploration and exploitation of mining deposits.

One of these techniques discovered by archaeological methods, is **the lithic industry or lithic technology,** that is, stone tools (different types of rocks and minerals),asopposed to metallurgy .

The archaeological find of the lithic industry, and the set of utensils that is its result, is a clear sample of humanactivity, although other animals(chimpanzees,, otters,, alimoches)occasionally use stones as tools.

The lithic industry in Prehistory comprises the following phases:

- **The Paleolithic** (pre-10,000 years AP) with lithic industry of boulders and silex objects .
- **The Mesolithic** (10,000 APs – 5000 APs), timed tools are manufactured (perforated, with saeta tips (peduncle tips and fins), with geometric microlithic tips (circle segments, trapezoids, triangles) and, above all, the production of small sheets that were fixed with resins to primitive snouts made of cane,, bone or wood .
- **The Neolithic** (5000 – 2000) with the use of silex, gold, silver and copper, which were perfecting as their intelligence and manual dexterity improved.

However, not only does the lithic industry provide us with first-born mininginformation, it also goes to the oil traits, as well as the remains of itselaboration, everything together within the archaeological register is a fundamental part of the knowledge of the human societies that manufactured it. These archaeological material remains provide us with important information such as:

- **Raw material** used, which tell us the sources of supply, the catchment area around the site and the displacements made by the human group.

- **Technique of elaboration** of the artifacts, which show us the technological development achieved at each time and their relationship with other contemporary human groups.

- **Spatial arrangement** of the tools within the site reveal to us workplaces.

- **Footprints of use and morphology,** which provide data on its usefulness or diet that its users followed.

THE OLDEST MINE

The oldest mine on an archaeological record and where we can check these **methods is the Lion Cave in Swaziland,** which according to the dating by **carbon method 14,has an age of** 43,000 years, the only one with neanderthal age. In this place, Paleolithic men dug for hematite with which they probably produced ochre-colored pigments. According to the National Trust Commission of Swaziland more than 1200 tons of hematite, rich in specular, were extracted from the Lion's Cave during the prehistoric era, 1200 tons with exploration, exploitation and use, without technology with a human force to date in anthropological study, we assume that Neanderthals had a higher strength compared to homo sapiens, according to recent studies.

An average Neanderthal could reach about 1.65 m, it was heavy in shape, prominent dentures and sturdy musculature, hovering around 70 kg in weight. This skeletal robustness produced an ability to support larger muscles, which thanks to its location to maximize lever action, gave the Neanderthal a greater physical force than Homo sapiens. Based on neanderthal skeletons found in the burials of *Shanidar (Iraq)*, this would certainly help faster artisanal extraction in greater quantities.),

The Ngwenya mines are located in the eastern foothills of the Drakensberg Mountains, near the northwestern border of Swaziland, in the Hhohho district. Mining activity was also practiced in other sites from the Stone Age, although more recently. The extractive industry, already defined as Homo sapiens, which took part in the original Ngwenya mine, did so contemporaneously with the presence of the extinct species of **Homo neanderthalensis on European** lands. The mining tools found in Ngwenya are of a recognizable type, specialized for the extractive purpose, and with characteristics typical of the site, different from those found in other archaeological sites of the Stone Age. In the main mining excavation of Ngwenya is first present mining technology that was used later in Europe.

This mining area shows the testimony of a mining cultural tradition that has disappeared, distinguishable by a wide use of social products made from the stone substance known as dolerita (diabasa). Such hachuelas, hammers and beaks, served for work on iron ore in order to extract red hematite (or red ochre), and specular (or specular hematite). Relatively isolated from other populations, the hunter-gatherer groups of those original inhabitants lost their cohesion 20000 years ago, their customs ceased to exist; its artists-magicians no longer used the ferrous resources used in the rituals of their spiritual tradition, and in cosmetic adornment. The reddish tone of the hematite operated, because of its similarity to blood, as a magic conferer of life, when applied on the bodies. The red ochre was also used, by the late peoples who gave rise to the present-day San (Bosquímanos), to concrete art on rock. There are a large number of these paintings in Swaziland.

GRIME'S GRAVES OF THE *NEOLITHIC*

Despite the evidence in the Paleolithic point to the use of lithic tools made of rocks or set of minerals for hunting development or household utensils, it is *in the Neolithic* that the traces of great mining found in many parts of the world appear, such as Belgium, France, Egypt, Spain, Germany and the United Kingdom, the most famous being **Grime's Graves located** in Norfolk England.

At first glance it looks like a field of small craters produced by explosives, like those of the battlefields of World War I. It is an archaeological site composed of 433 mining wells built in Neolithic times to reach the coveted veins of silex.

The silex was one of the most coveted and valuable materials of the Stone Age, used for the manufacture of cutting tools and weapons. In addition to its hardness and ability to easily break into sharp edge sheets, hitting it against other rocks produced sparks, ideal for making fire.

Our ancestors soon discovered these qualities and already from the Paleolithic organized their extraction from the subsoil through wells and tunnels, a mineractivity that had its greatest development during the Neolithic. Researchers have determined that they were exploited between about 3000 and 1900 BC. The wells extend over an area of 37 hectares and the largest reach more than 14 meters deep and 12 meters in diameter on the surface, without the great technologies developed in recent centuries, by Neanderthal force. It is estimated that up to 2,000 tons of limestone had to be extracted in Grime's Graves to create the largest wells, which would have required about 20 people to work for five months.

Extrapolating the data in Grime's Graves tells us that between 16,000 and 18,000 tons of syllex could be extracted, which would serve to produce some 3 million tools, axes and other artifacts. Much of that raw raw material would probably be used for trading, and would be carved in certainly remote locations. With this safe pace one or two of the wells were exploited at the same time, opening new ones every one or two years and filling the previous ones with the land and rocks of the latter.

The site consists of three layers or veins of silex, which were exploited successively while excavating the wells, the last and deepest being the richest of all. In the extraction work they would have used wooden platforms and stairs.

This is known from subsequent studies that have carried out some 28 wells excavated until 2008, where several hundred tools made of deer antler have been found that miners used as beaks. Probably the complement were wooden shovels, as remains of them have been found in other deposits. Once the vein was reached vertically, something that is already really impressive using deer antler velvet, horizontal galleries were dug following it to get as much extracted as possible

In modern times its function and purpose were not discovered until in 1870 archaeologist William Greenwell dug one of the wells (the same Greenwell that two decades later would find folkton's famous and enigmatic Drums).

These English never imagined that they would be the pioneers in mining, from an element as basic as Silex, but as useful for survival and development of today's societies.

This aroused in humans the search for other natural resources as useful as Silex but which we have turned into conflicting or lethal minerals in some cases, in both cases resulted in the search for complex mining such as Tanzanite, Gold, Copper, and how conflicting some deposits such as Coltan and Wolframita would become, and how harmful the uraninite radioactive groups or the carnotite, it's amazing how we went from being helpless beings looking for tools to hunt and give a daily livelihood home to using our resources for our own destruction, whether it's for war finance, personal interest, or just wanting to have a recognition of being a power among other countries.

METALWORKING *METHODS IN ANCIENT EGYPT*

In Ancient Egypt, the mining, refinery and metalworking techniques began during the early dynasties, where its inhabitants extracted malachite as well as imported silver, copper, tin and lead elements that were used for ornamentations, ceramics, jewelry and decorations.

The ancient Egyptians used their expertise to survey ore ores in Egypt and other countries. Ancient Egypt had the means and knowledge to carry out the surveys of ore ores they needed, establish mining processes and transport heavy loads over long distances, by land and sea.

The Egyptians possessed important knowledge of chemistry and the use of metal oxides, as manifested in their ability to produce glass and porcelain in a variety of natural colors. In addition, the ancient Egyptians made precious colors reflecting their knowledge of the composition of different metals, and knowledge of the effects produced by the salts of the earth on different substances. This matches our "modern" definition of chemistry and metallurgy.

The metalworking methods: casting, forging, welding and engraving of metals, not only was much practiced, but were also the most developed. Frequent references in the steel registers of Ancient Egypt give us a more real conception of the importance of this industry in Ancient Egypt.

The Egyptians' ability to prepare metals is more than proven with vessels, mirrors and bronze utensils, discovered in Luxor (Thebes), and elsewhere in Egypt. They adopted numerous methods to change the composition of bronze, through a successful combination of alloys. They also had the secret of giving a certain degree of elasticity to bronze, or brass leaves, as evidenced in the dagger that houses the Berlin Museum today. This dagger stands out for the elasticity of its leaf, its neatness and the perfection of its finish. Many products from Ancient Egypt, which are currently spread by European museums, contain between 10 and 20 parts of tin, and between 80 and 90 parts of copper.

His knowledge of metal ductility is evident in his ability to produce wires and metal wires. Wire drawing was achieved with ductile metals such as gold and silver, as well as brass and iron. The thread and the gold wire were the result of wire drawing, and there is no case of them smoothing.

The Egyptians perfected the art of making the metal thread. This was thin enough to intertwine in the fabric, and for ornamentation. There is a certain delicate linen of Pharaoh *Amosis,* with numerous animal figures working with gold threads, which require a high degree of detail and delicacy.

The science and technology to manufacture metal products and goods was known and perfected in Ancient Egypt, it is known that more than 5,000 years ago they could produce numerous metal alloys in large quantities. The ore ores that Ancient Egypt lacked to produce copper and bronze alloys (copper, arsenic and tin) were obtained abroad.

The three ore ores were only imported from the known source in the ancient world, Iberia. Archaeological records show the ancient use of the mineral resources of southern Iberia of copper and arsenic. As for the tin, we know the "Tin Route", which ran along the western coast of the Iberian Peninsula, where the tin arrived from Galicia and possibly from Cornwall

Egyptian copper was hardened by the incorporation of arsenic. Variation in composition has been observed: for example, daggers and wirebars had stronger sharp edges, and contained 4% arsenical copper, while axes and tips contained 2% arsenical copper.

In addition to making arsenical copper, the ancient Egyptians also made bronze products. The incorporation of a small ratio of tin into copper produces bronze, resulting in a lower melting point, increased hardness and easier casting. Tin content varies greatly between 0.1% and 10% or more. Many bronze objects from very remote periods have been found. The bronze industry was very important to the country. Bronze was perfected and used in Egypt for large vessels, as well as tools and weapons.

The Egyptians used various types of bronze alloys, as we know from the texts of the New Kingdom, where there is a frequent mention of "black bronze" and "bronze in the combination of six", i.e. a six-component alloy. These variations produced different colors. Yellow brass was a compound of zinc and copper. A type of white brass (and thinner) had a silver mixture, which was used for mirrors, and was also known as "Corinthian brass". Incorporating copper into the compound produced a yellow color, almost a golden appearance.

Copper and bronze produced material for a wide variety of household utensils, such as cauldrons, jugs, buckets and a wide range of tools and weapons, such as daggers, swords, spears and war axes.

But not only were they experts in the exploration and exploitation of metallic minerals, we know that these techniques also applied them for other non-metallic minerals.

One of these productions focused on glass articles, from the pre-nautical period. The stained objects of this ancient era were mainly pearling, with solid quartz or steate to be used as a nucleus. The steatita was used to carve small objects such as amulets, pendants and *small neteru* figures (gods/goddesses), as well as a few larger items, and offered an ideal base for glazing. The stained-glass objects were discovered in the dynastic period [3050–343 BC], and is by far the most common material for beetles. The same glazing techniques were used to mass produce burial equipment (amulets, *shabti figures*) and domestic decoration (blues, inlays with floral motifs).

The high quality and wide variety of glass articles of Ancient Egypt are indicative of knowledge of the metallurgy of Ancient Egypt. The most common colors of Egyptian glass were blue, green or bluish green. Color is the result of adding a copper or malachite compound, which by the time was very common to use. The brightest results were achieved by using a mixture of copper and silver. The range of colors of these semiprecious stones is fascinating, ranging from the clear blue of lapis lazuli to the turbulent blue of turquoise and the speckled gold of cornaline, these are the three most representative stones of the art of Egyptian jewelry. However, there was also agate, amethyst and hematite. In addition, we should keep in mind that the Egyptian craftsman did wonders with enamel, large plates that were decorated with hieroglyphics or gussets.

Ancient Egyptian crystal was formed by intensely heating quartz sand and natron with a small mixture of coloring agents such as a copper compound or malachite, to produce both green and blue crystals. Cobalt was also used, which had to be imported. Then the ingredients melted into a molten dough, the heating ended when the dough achieved the desired properties. With the dough cooled, it poured into molds, and spread into thin rods or reeds or other desired shape.

The ancient Egyptians showed their excellent knowledge of the different properties of materials in the art of staining glass with different colors, as is apparent from the numerous fragments found in the tombs of Luxor (Thebes). Their skill in this complicated process allowed them to imitate the intense brilliance of gemstones. Some imitation pearls have been so well imitated, that even today it is difficult to differentiate them from real pearls with powerful lenses.

Finally, the Egyptian mining experience is ordered in the civilization of Ancient Egypt, where they kept written records showing the nature of their expeditions and the preparations for their mining activities. Records from Ancient Egypt show an impressive organization of mining activities more than 5,000 years ago, in numerous locations both outside and within Egypt. The turquoise mines *of Serabit el-Jadim* on the Sinai Peninsula show a typical ancient Egyptian mining quarry, consisting of a network of caverns and carefully excavated horizontal and vertical passages with appropriate corners, as were the quarries of ancient Egyptians in all periods. Ancient Egyptians could dig long and deep into the mountains with proper propping and support from excavated wells and galleries. Groundwater infiltration into galleries and wells was safely pumped to ground level. These Egyptian bombs were famous all over the world, and were used in Iberia's mining activities.

In the *mines of Uadi Maghara,* in Sinai, the stone huts of the workers still remain, as well as a small fortress, built to protect the Egyptians located there from the attacks of the Bedouins of Sinai. There was a well of water not far from these mines, and large cisterns in the fortress to keep the water. The mines of Uadi Maghara were active throughout the dynastic era [3050–343 BC].

In addition to this the Egyptians were very religious always associated their creativities with the construction of temples/sanctuaries along with commemorative stees near/at each mining site. The exact same practices were found in mining sites outside Egypt, such as in the Iberian Peninsula, where it was extracted from silver, copper, etc. mines since time immemorial

Each mining site was conceived and planned with development plans. Two papyrus from Ancient Egypt were found, including maps of places, related to gold mining activity during the reigns of *Pharaohs Seti I and Ramesses II*. A papyrus, which is only partially preserved, represents the golden district of Bechen Mountain in the Arabian Desert, and belongs to the time of Ramesses II. The plane of the place of the papyrus found represents two valleys that run parallel to each other between the mountains. One of these valleys, like many of the largest valleys in the desert, is covered in undergrowth and stone blocks that control soil erosion as a result of surface water evacuation. The site-ready plan shows the main details of the site, such as the road network within the mining site and its connection to the outer road system and the "routes leading to the sea". The plan of the place also shows treatment areas of ore metals, small houses, storage areas, several buildings, a small temple, a water tank, among other things. The surrounding area of the mining site shows cultivated land, to supply the food needed for the colony of the mining site.

Records from Ancient Egypt show the organizational structure of mining operations. Records preserved from Ancient Egypt show the names and titles of several officials who, during the Ancient and Middle Kingdoms, led the works *in Hammama*t, in *the Bechen mines in the Arabian Desert*. These included engineers, miners, blacksmiths, masons, architects, artists, safety detachments and boat captains, who maintain the integrity of the boat parts for reassembling when the expedition manages to reach navigable waters.

Ancient Egyptians searched for raw materials in other countries and used their indigenous experience to exploit, extract and transport raw materials from all over the inhabited world.

THE MINERIA IN THE RISE OF THE ROMAN EMPIRE

In *roman times,* ***the*** **mining industry** in **Europe** boomed significantly. The Romans contacted a Mediterranean country rich in mines, when in 206 a. C. the Carthaginians of Hispania were thrown on the occasion of the Second Public War. The Iberian Peninsula was the richest district in the entire Ancient World and the first to be exploited by the Romans on a large scale, were an increase in the mineral sources of Rome, allowed by the technology and metallurgy of the time.

Hispania, which already had an important mining tradition, gradually becomes, during the period from 218 BC to 19 BC (end of the Cantabrian Wars), an authentic mining colony for Rome.

Ancient sources recount Hispania's great mining riches. Mela (II, 86) and Pliny (NH. III, 30), among others, claim that the most abundant minerals are iron, lead, copper, silver and gold. Estrabón (III, 2, 9) alludes to the richness of silver, tin and white gold, mixed with silver, contained in the northwest territories of Hispania. Riches that undoubtedly helped the rise and success of the entire Roman Empire, with this mixture of riches added to the alloy of metallic minerals to build stronger and lighter weapons in the relationship with existing ones of the time, they turned them into an imposing and lethal army leading them to occupy large areas of land and this in turn generated a growth of society creating a thirsty demand for natural resources to meet the needs that would enable a great development and sustainability of the empire.

Not everything was gold color, the Romans went through great challenges that led them to solve some serious inconveniences that presented themselves day by day incorporating in the planning ideas to concentrate technology and Innovation, which at the end of the day gave the necessary development to make a recovery more efficient, although it would be daring to ensure that safer, which is possible in our day, the combination of productive and safe company where we have the great opportunity to be our own sustainable Roman titans.

One of the great problems that the Romans were able to solve, unlike their predecessors, was the water problem. In an exploitation it is necessary to deepen more and more to be able to extract the material, keeping the production levels alive. As water deepens in this exploitation, water increases. Thanks to the topographical knowledge of this practical village and its water extraction machinery it was possible to solve this problem.

The great geological knowledge and skill in the field surveying work made mining exploration very thorough and intense. So much so that it is now still a source of astonishment. However, some experts say that in the gold prospecting work of the Romans they are based on a systematic application of empirical criteria.

In underground excavations, the lighting of the galleries was carried out by oil lamps made of cooked clay, placed in small cavities, which are called skylights, excavated at the height they deemed appropriate.

Iron tools are introduced for material removal, allowing optimization of work performances, no longer using stone or bone tools. The techniques of fire and water to work with very hard rocks are still used at this time, both in the advancement in galleries and for the despondenging of rocky masses. This method may have some limitations in small or low-ventilation environments.

It emphasizes the use of wood for the maintenance and lying work, previously used by other peoples. The use of this material has some advantages, such as being abundant and easy to work with. Later, they resorted to using rock as a complement to wood. In transport, the Romans carried out the lysing operations using cables made of vegetable or leather fibers, and sometimes used them with a pulley system.

For drainage operations, the use of the Ferris wheel or bucket wheel stands out. These were driven by the force of man, who had to step on some crossbars located on the outside. Some wheels found reach a diameter of 4.5 meters. Other systems used were Archimedes Screw or Double Piston Pumps.

The Romans' contribution to open pit mining was the use of hydraulic force for gold mining. Water was used for both washing and material extraction, which implies a reduction in labour. The most impressive example of open pit Roman mining can be found *in Médulas de León,* with the construction of a hydraulic network with 600 kilometers of canals.

All this data magnifies the planning and structuring of mining work carried out by the Romans, making the exploitation of existing deposits much more fruitful than previous peoples had done. Without fail to appreciate the legacy that leaves us, and the historical relationship of how societies adapt to the needs existing at a characteristic moment of the time, which in some cases marks a milestone in contemporary history, that speed of adaptation implies a benefit translated into development and well-being.

URBAN DEVELOPMENT IN *ANCIENT GREECE*

During *Ancient Greece,* a variety of minerals and precious stones were extracted for the construction of palaces, temples and sculptures. Much of the Greek practices were adopted by the Romans, although these contributed to the industry from the construction of **aqueducts** that allowed them to find and exploit new minerals.

In the 4th century BC, *Tóricos and Laurion,* which is next door, became the most important mining district in Greece, thanks to new technical procedures to extract silver

One of the highlights of Ancient Greece is *the Mycenaean Acropolis of Tortoica,* a fortified citadel that dominates the natural port of Lavrio, south of the Attica region. A strategic site not only because of what was collected from the site, which was naturally protected by a small island called Macri, but also because minerals such as silver and lead were extracted from there.

The galleries are developed by several overlapping levels and their structure allows to imagine what the evolution of mining activity was like over time. Where they discovered axes connecting the two main levels despite the difficulty of access, where it clearly forces the use of **climbing and caving techniques.**

The discovery of ceramics and volcanic stone hammers, manufactured in a nearby quarry, is a reference for chronological dating between **the final stage of the Neolithic** and the initial of the **Heladic,** about 3200 BC. This date, if confirmed by complementary research, would revolutionize those known so far for mining activity in the Aegean area.

The Laurion mines are copper and lead mines, but mainly known for the silver metal they produce. Many remains (wells, galleries, surface workshops) still mark the landscape of the southern tip of Attica, betweenO tórico and Cape Sououunión, about fifty kilometers south of Athens,, Greece .

After a prehistoric phase of copper and silver galena exploitation, an overall recovery of the holding dates back to the classical period.

The Athenians deployed spectacular and inventive energy to make the most of the mineral, affecting in particular many slaves. This contributed significantly to the city's fortunes and was an undoubtedly decisive factor in the establishment, on the scale of the Aegean world, of Athenian thalassocracy. The development of the Athenian currency and its function as a reference currency throughout the Greek world at the time also explain the wealth of deposits exploited in Laurion, the first major milestone in the history of silver mining.

In the 5th centuries a. C.A. and IV to. C., the Athenian city earned significant income from the exploitation of silver. His real discovery probably skidds to the last quarter of the 6th century. C.

The benefits from the exploitation of the Laurion mines greatly contributed to sustaining Athens' imperialist policy in the 5th century BC. , and these benefits would have looked at outside the region which explains Sparta's willingness to hinder the extraction of metal resources, and initiate systematic attacks which would lead to the beginning of the Peloponnesus War, through militaryincursions into the Attica region and especially in Laurion, with the aim of devastated the production infrastructure, as the Spartans were aware that the war did not depend so much on weapons and men, as on the money that allowed its large-scale manufacture and maintenance.

As in today's times, interests overlap, causing thoughts of greed between regions, distinguishing the way to a bilateral understanding, which well planned as the Romans did would give benefits in all directions.

Another disadvantage of lack of cooperation is the resumption of mine activity, where it was slow and progressive, at least until the first third of the 4th century BC. Small amounts of ore were exploited in surface workshops and old galleries, without opening other new galleries. Such a lack of investment is so explained by the difficulty of reconstitution of the large existing workforce before 413 a. C. and for the low income earned by the Athenians from the concessions of mine mining, in relation to the amount of money they invested to open new galleries.

All this leads to a considerable delay in exploitation and it was not until the second half of the 4th century a. C., when the mines reached the highest levels of production, the prospecting and opening of new wells and galleries multiplied, corroborated by the fact that most of the lamps found belong to this period.

That uptick in production was due to the great expertise of Laurion miners who were far from acting randomly. They had acquired a very precise knowledge of the geological characteristics of the subsoil and applied their theoretical knowledge to direct their research, they dug there because they knew the special conditions for obtaining marble for example, they knew that there should be a bottom and a cap so they began to investigate where logically there should be ore, and what those were special conditions to identify the productive area, again innovation helps us to improve recovery.

One of these manifestations of sequential stratigraphy is *the Kitso*well, in *the Maronea region,*the miners began their exploration in the upper marble, understood that it was thin; five meters below, they reached the shale,without even taking an interest in the second contact, continued their descent to a layer of marble, 59 meters deep. Thinking that they were at the level of the third contact (at the lower limit of the shale layer and upper of the second layer of marble), rich in ore, they dug side galleries, without finding the desired mineral. In fact, this layer of marble was actually a thin block of limestone inserted into the shale, the actual contact was twenty meters lower. This is a rare case in Laurion, and therefore did not correspond to the science of miners. They concluded, based on their knowledge, the absence of ore there and therefore abandoned this fruitless investigation.

In this way the mining farms of Laurión have the deepest vertical wells of antiquity. From a rectangular or square section less than two meters wide, sometimes descend to more than one hundred meters (119 meters the deepest known) but more generally between fifty and sixty meters. They are cut very evenly, so that each side is flat. Its verticality is surprising: It is estimated that two workers took twenty months to excavate a well 100 meters deep.

The galleries were narrow (50 to 60 cm wide, 60 to 90 cm high), which did not facilitate the work and movement of the miners or the evacuation of the extracted ore, which leads us to think that the work was done by a single person. When the excavations were larger, workers left portions of rocks poor in minerals that served as pillars (ormos), the narrowness of the galleries allowed to limit in most cases the risk of collapse; it also accelerated progression. The miners attacked the cutting front by excavating notches 12 cm wide throughout the gallery height. After five notches, the entire gallery, 60 cm wide, had progressed from the depths of these notches. The work lasted about ten hours, which corresponds to the speed of rotation of the equipment and the lighting capacity of the lamps, afterone month, the galleyhadbeen excavatedten meters.

All this movement was made with the help of a hammer (tukos) of 2.5 kg of short handle (twenty to thirty centimeters) of olive wood and whose head had a four-sided tip to break the rock on one side and a flat head on the other. This flat head was used to hit double-sided chisels or metal rods 2–3 cm in diameter called quarry pointers (xois), 25–30 cm long and with a sharp four-sided beveled end. Given the very harsh nature of the marble in which the galleries were excavated, it is estimated that a worker had to use between ten and thirteen tips in ten working hours, tools that he had to repair and sharpen regularly. This is despite the fact that these hammered and tempered iron tools were made of an excellent quality metal. The beak, which usually consists of a four-sided tip on one side, a hammer capable of nailing at one point or corner on the other, is the third basic tool. Several copies of these abandoned tools have been found in the galleries. It was generally used to attack the most fragile parts of the rock.

Workers also used iron hooks to collect debris collected in esparto baskets or leather that another slave dragged through the well, from where they were raised by a pulley system. Today, we can see on the surface the remains of the low walls that were used as crane anchors.

The miners used small terracotta candles for lighting, identical to those commonly used by the Greeks of the time in their daily activities, which smoked a lot and consumed a portion of the scarce oxygen from the air, which became an important point to solve. They usually have a single nozzle, but have been found with several nozzles, and then used to illuminate important intersections or large construction works. At its end a new team took over the operation of the mine's day and night.

It seems incredible that all this machinery was coupled, in the first instance, to support the troops, it was important to support the army with all its clothing plus the utensils they needed in the fields of war, in addition to the minting of money to sustain the empire in constant economic movement.

During the first medieval centuries, the departure of metal was in a steady decline and restriction on small-scale activities were the main cause for economic, political and social stagnation in the decline that followed the Roman world affected by Europe, throughout the early medieval period, and had a fundamental impact on technological progress, trade and social organization. Technological developments that affected the course of metal production.

THE MINERIA, THE ADVANCE IN *THE MIDDLE AGES*

In the Middle, economic and social conditions pushed the sector to focus mainly on copper extraction, iron and other precious metals, using them to mint coins and warrior clothing: weapons, catapults, etc., and dictated the increased need for metal for agriculture, arms, stirrups, and decoration, this began to favor metallurgy and slow but steady overall progress was observed, which rebounded with the discovery of the new world.

The period immediately after the twentieth CENTURYmarks the widespread application of several innovations in the field of mining and mineral treatment. It marks a large-scale change and better quality of production. medieval miners and metallurgicals had to find solutions to the practical problems that limited previous metal production, in order to meet the demands of the metals market.

Thanks to high demand for silver and copper, mining activity in medieval Europe became the most important economic sector, after agriculture.
While it is true that demand for oriental products such as silk and spices was high, these in Europe did not have many luxury or high value products to offer and the export of raw materials, in general, was not profitable because of the high freight costs. It was mining products such as silver, copper, bronze and recent brass that constituted the main European objects.

Brass was discovered by the Romans, but initially its use was very restricted by the complex and costly process. With the rise of European cities, demand for this alloy also increased, lending itself especially to decorative and religious objects. The first brass productions in the Middle Ages date back to the 10th century in Dinant, Belgium, an area rich in calamine, a zinc carbonate used for the manufacture of brass. This zinc and lead metal oil area spread to Aachen and Cologne in Germany.

Dinant's artisans quickly took over the market and their products became famous throughout Europe. The introduction of hydraulic hammers (15th century) in the grinding of ore and metal forging greatly increased cost efficiency. Copper for brass manufacturing came from Harz (Germany) and Falun (Sweden).

The powerful guilds were very jealous in protecting their craft as metallurgical and did not push away conflicts and wars with rival cities. To protect their craftsmen, they even banned the use of modern hydraulic hammers in metal work, which was not the case for rival cities such as Aachen and Stolberg in Germany. The consequence was that Dinant could no longer compete, his best teachers emigrated and created in Stolberg the new center of brass in Europe. Opposition to technical advances eventually ended Dinant's four-centuries leadership in the world of brass.

Silver and copper were mainly produced in Germany, tin came from Wales and brass was manufactured in Belgium, and then in Germany. Gold came from West Africa (Ghana, Mali) and reached the Mediterranean through caravans that crossed the Sahara Desert via the Sudanese route. This monopoly of gold was one of the factors that the Portuguese sought to break (outside the commercial control of spices) in their quest to seek a sea route to the East. In their maritime discoveries in the mid-15th century they reached the coasts of Mali and the Kingdom of Akan (Ghana) and began marketing copper, brass and glass against gold and slaves. Since maritime transport was much cheaper than the cost of caravans that required months to pass through the thousands of kilometres needed to reach the Mediterranean, control of the gold trade quickly passed into Portuguese hands.

In the Middle Ages all mines belonged to the king or emperor, who temporarily granted them to private entrepreneurs against a particular monthly payment. This lease had to be less than the profit left by the business and the profit depended primarily on the price. Well, the easiest way to raise prices was to create a monopoly. Said and done: Charles V ceded to the Fuggers all copper, silver and mercury mines in his vast empire (excluding America), which allowed him to drastically increase the value of the concession and according to it the price was set. This brought huge amounts of money into the royal and imperial coffers, eventually allowing Charles V to repay the large sums owed.

At all stages of humanity, we see mining as a fundamental pillar of development for our societies, from the Paleolithic, through the Mesolithic to the Neolithic, man always look for the natural resources that our planet gives us to meet our needs, in some cases personal needs.

In any case, the impeccable search for resources has led us to apply new technologies and innovate in the exploration and exploitation phases to make the best use of minerals, we go from lithic tools to the development of large refineries and alloys to obtain new materials diversifying resources.

In our time the development of technology is thanks to scientific research, which has to lead us to new environmental planning, it is true that our resources are necessary for the development and maintenance of societies, but it is also true that research must make these resources longer by giving a ratio of lubricants – materials – consumption to decrease friction and be able to have high quality products avoiding wear and tear. Now it's up to us to use those same tools to educate on the efficient and responsible use of all equipment, utensils, tools, means of transport, and everything involving minerals or natural resources.

2 BRIEF HISTORY OF THE OIL INDUSTRY

Oil use has been known since prehistory, but with that said we have known it since the 19th century. The indigenous people of the pre-Columbian era in America knew and used oil, which served as a waterproofer for boats, a process that served for the perfection of European ships.

For several centuries the Chinese used petroleum gas forcooking food, bonfires to keep the cold away, among other things. Before the second half of the 18th century, however, the applications given to oil were very few.

It was Colonel Edwin L. Drake who drilled the world's first oil well in 1859, in the United States, managing to extract oil from a depth of 21 meters.

It was also Drake who helped create a market for oil by separating kerosine from it. This product replaced the whale oil used at the time as fuel in the lamps, the consumption of which was causing the disappearance of these animals, marks the replacement of one resource for another.

But it was not until 1895, with the appearance of the first cars, that gasoline was needed, that new fuel that in the years that would be consumed in large quantities. On the eve of World War I, before 1914, more than a million vehicles using gasoline already existed worldwide.

Indeed, the true proliferation of automobiles began when Henry Ford launched his famous "T" model in 1922. That year there were 18 million cars; by 1938 the number rose to 40 million, in 1956 to 100 million, and to more than 170 million by 1964. Today it is very difficult to accurately estimate how many hundreds of millions of gasoline vehicles exist in the world.

Logically, crude oil consumption to meet gasoline demand has grown in the same proportion. It is said that in the 1957 to 1966 the amount of oil was used in almost the same amount of oil as in the previous 100 years. These estimates also take into account the expense of aircraft with piston engines.

Another fraction of crude oil served as energy is diesel, which before 1910 was part of the heavy oils that constituted refinery waste. Consumption of diesel as fuel began in 1910 when Admiral Fisher of the British fleet ordered coal to be replaced by diesel on all its ships. The best argument for making such a decision was its heat superiority over mineral coal.

Later the use of this energy was extended in the merchant navy, in steam generators, in industrial furnaces and in-home heating.

The use of diesel spread rapidly to Diesel engines. Although Rudolph Diesel invented the engine that bears his name, shortly after the internal combustion engine was developed, its application was not very successful as it was originally designed to work with pulverized coal.

When the light fraction of diesel, which was called Diesel, was finally separated, Rudolph Diesel's engine began to find ample development. Therefore, these engines found rapid application on military and merchant navy ships, railway locomotives, heavy trucks, and agricultural tractors.

After this brief analysis of the history of the development and use of petroleum fuels, we clearly see that the largest consumer of these energy is the automobile.

This is due not only to the fact that millions of vehicles with internal combustion engines are in circulation, but also because of the very low efficiency of their engines, as they would waste 75% of the energy generated, as mentioned above.

Thus, as the automobile remains the main customer for most oil refineries that are designed to provide gasoline to this closed market.

After the appearance of the car, the world began to move faster and faster, requiring day by day higher power vehicles, and therefore better gasolines.

OIL FORMATION

There are several theories about oil formation. However, the most accepted is the organic theory that is assumed to have originated from the decomposition of the remains of animals and microscopic algae accumulated at the bottom of the lagoons and in the lower course of the rivers.

This organic matter was gradually covered with increasingly thick layers of sediment, sheltered from which, under certain conditions of pressure, temperature and time, it was slowly transformed into hydrocarbons (compounds formed of coal and hydrogen), with small amounts of sulfur, oxygen, nitrogen, and traces of metals such as iron, chromium, nickel and vanadium, the mixture of which constitutes crude oil.

These conclusions are based on the location of oil mantle, as they are all found in sedimentary land. In addition, the compounds that form the aforementioned elements are characteristic of living organisms.

Now, there are people who don't accept this theory. Its main argument lies in the inexplicable fact that, if it is true that there are more than 30,000 oil fields worldwide, so far only 33 of them constitute large deposits. Of these large deposits 25 are located in the Middle East and contain more than 60% of our planet's proven reserves, in all deposits only approximately 30% of their totality can be recovered with the best of technologies.

Other theories that argue that oil is of inorganic or mineral origin. Soviet scientists are the ones who have been most concerned with testing this hypothesis. However, these proposals have also not been fully accepted.

An interesting version of this topic is the one published by Thomas Gold in 1986. This European scientist says that natural gas (methane) that is usually found in large quantities in oil fields could have been generated from meteorites that fell during Earth formation millions of years ago.

The arguments it presents are based on the fact that more than 40 querogen-like chemicals have been found in several meteorites, which is supposed to be the precursor to oil.

And since NASA's latest discoveries have proven that the atmospheres of the other planets have a high methane content, it's no wonder that this theory is gaining more and more followers every day.

We can conclude that despite the countless research that has been carried out, there is no foolproof theory that explains without a doubt the origin of oil because this would mean being able to discover the origins of life itself.

This viscous liquid whose color varies from dark yellow and brown to black, with green reflections with a characteristic odor and floats in the water. It is the subject of great controversy and ambition for everything it generates on the planet.

But if you want to know everything that can be done with oil, this definition is not enough. It is necessary to deepen knowledge to determine not only its physical properties but also the chemical properties of its components.

The mixture of hydrocarbons, compounds contained in theirmolecular structure carbon and hydrogen mainly, where the number of carbon atoms and the way they are placed within the molecules of the different compounds provides oil with different physical and chemical properties. So, we have that hydrocarbons composed of one to four carbon atoms are gaseous, those containing 5 to 20 are liquid, and those of more than 20 are solid at room temperature.

Crude oil varies greatly in its composition, which depends on the type of deposit from which it comes from, but on average we can consider that it contains between 83 and 86% carbon and between 11 and 13% hydrogen.

The higher the carbon content relative to hydrogen, the higher the amount of heavy products that crude oil has. This depends on the age and some characteristics of the deposits. However, it has been proven that the older they are, the more gaseous and solid hydrocarbons and the less liquid they enter their composition.

The composition of crude oil also contains sulfur derivatives (which smell like rotten egg), in addition to carbon and hydrogen.

In addition, crude oils have small amounts, in the order of parts per million, of compounds with nitrogen atoms, or of metals such as iron, nickel, chromium, vanadium, and cobalt.

Oil as it is extracted from wells generally does not serve as energy as it requires high temperatures to burn, as crude oil itself is composed of hydrocarbons of more than five carbon atoms, i.e. liquid hydrocarbons. Therefore, in order to be able to take advantage of it as an energy it is necessary to separate it into different fractions that constitute the different fuels. And all this requires the development of technologies worthy of such an enigmatic and prolified natural process as the formation of Petroleum.

PRE-COLUMBIAN OIL

Oil and its relationship with men is more varied and earlier than is believed. In different parts of the world he was known by different names. Bitumen, naphtha, and Tar, in the ancient cultures of the Middle East, rich in natural manaderos of this substance.

In America it has several names recognized by *archaeologists, the splash of* the ancient Mexicans, the *copé or copey,* word of origin tallen, a people that flourished in what is now northern Peru, *and Mene* on Venezuelan soil, which, if it is clear that since their unforeseen departure to our surface, was regarded by our Aborigines as the curator of diseases or any other physical condition, for they knew of your chemical properties and regarded them as a blessing bestowed upon them by the Gods, of their religious beliefs. With the thickness that characterizes you, it can be made up of more than one chemical compound that allows the creation of many products or their product.

As for the uses that the ancient man found for this substance, one of the first and most obvious was the use of its combustible qualities to produce light, using hollow reeds or candiles, along with auquénid wool wicks or plant fibers. The wit invented in remote age produced a bright light, but it smoked and tized heavily around it. Another use, this time military, that was found for the fiery tar was to use it in the incendiary arrows, a very powerful weapon when it came to besieging some city or building. There is another utility that Andean cultures found for this substance. Without contact with fire, tars heated in clay containers could serve as binders. Although archaeologists still dispute this, it is common ground in the written testimonies that the Inca roads, at least in certain sections, had already used a class of asphalt for their construction. The most detailed at this point has been the American historian William Prescott, who, quoting Sarmiento de Gamboa and Garcilaso de la Vega, described an Inca path composed "of large stone steeds, covered at least in some parts, with a bituminous mixture to which time had given a hardness greater than that of the stone itself".

Also, with natural tares would be prepared a bitumen that, was applied on the face in some religious rites. In the vicinity to other natural suppliers, in the jungle, for example, their virtues as insect repellent, would hardly have gone unnoticed. This observation, along with another on a use against the mites of the Auquénids in the pirín manaderos, in the basin of Lake Titicaca, has been made by the historian Waldemar Espinoza. For his part, antonio Raimondi had already observed in the nineteenth century, the curious spectacle of a pig trope stirring instinctively in this "fat of the earth".

To these ancestral and home uses, always practiced on a small scale during pre-Columbian times that its use dates back approximately 2 thousand years old, by the first Mesoamericans, two practical applications of importance during the three colonial centuries were added. With them a new phase was inaugurated in the exploitation of the black, viscous and oily substance, a more intensive but certainly insignificant exploitation compared to the scale of the farm that would come next.

The first of these uses was to use tars as paint on woods that were periodically to be made to ships. In addition to better waterproofing against salt water, tar caulking protected keels for some time against one of the most terrible enemies of sailors, the prank or seaworm, which was eating the bottoms of a wooden vessel, this was crucial to sealing and making boats more efficient.

It was a highly appreciated use and those who know it vital for maritime transport, but the most intensive use, the tars acted in this case as waterproofers in the bottoms of the mud boots that were used to package and transport any drink that had prepared. Oil was also a very early introduction to colonial agriculture and the production of wines and spirits was already a significant event at the end of the 16th century.

There were other practical purposes for oil or "stone oil, as it was called, although our mentality of modern men would hardly consider them. Such was the case with the alleged medicinal applications that some believed had tars. According to historian Pablo Macera, he claimed that oil served to combat "poisoning, nerve flojera, uterine suffocation, verminose effects, menstruo suppression and tumors." Conquest and the three colonial centuries found a wide variety of uses to this substance, but the hidden powers of oil were still unknown.

During the colony, and even for many decades of the Republic, nothing significant happened in terms of the use of this substance. Technical developments or even energy needs did not press hard enough, as they would at a later time, when machinism and industry would change the face of the earth. Then the hour of oil would sound, and their search and extraction would reach magnitudes that the ancient settlers would never have been able to dream of.

ASIA AND EUROPE WITH THE FIRST HYDROCARBON DEPOSITS

Between 6,000 and 2,000 BC, the first gas deposits were discovered in Iran, which served to feed the *"eternal fires"* of the fire worshippers of ancient Persia.

According to Herodotus and the confirmation Diodorus Sículo, asphalt was used in the construction of babylonian walls and towers, there were oil wells in Arderica (near Babylon) and a source of tar in Zante ((Ionian Islands).). Large quantities were located on the banks of the Pinarus River, one of the tributaries of the Uphrates. Tablets from the ancient Persian Empire indicate the use of oil for medicinal and lighting purposes in the upper classes of society.

Oil was exploited in the ancient Roman province of Dacia, present-day Romania. According to Dioscorides, oil floating in springs in Agrigento was used in lamps instead of olive oil.

The first streets of Baghdad were paved with tar, derived from oil that was naturally obtained from the region's fields. In the NINTHcentury, oil fields were exploited around modern Baku,, Azerbaijan. These fields were described by the Arab geographer Al-Masudi in the 10th century, and by Marco Polo in the 13THcentury, who quantified the production of wells as the shipment of hundreds of ships. Chemicals such as Kerosene were obtained in still (al-ambiq) for use in lamps. Thatmicos arabes and Persians alsodistilled rawpetrorleo in order to obtain flammableproducts for military use. Through Islamic Spain, distillationbecame known in Western Europe in the 12th century. He was also present in Romania from the 12th century, becoming known as a pocur.

In 1710 the Swiss physician, and master of Greek, Eirini d'Eyrinys (also written as Eirini d'Eirinis) discovered asphalt in Val-de-Travers,, ((Neuchatel). He established the Presta bitumen mine there in 1719;which was operational until 1986.

In 1745 under the reign of Empress Elizabeth I of Russia,, Fiodor Priadunov built the first oil and refinery well in Ukhta. By distilling "rock oil"(oil)it obtained a kerosene-like substance used in oil lamps in Russian churches and monasteries.

Bituminous sands were exploited since 1745 in Merkwiller-Pechelbronn,, Alsace,under thedirection of Louis Pierre Ancillon de la Sablonniére,under special mandate of Louis XV of France.. ThePechelbronn oil field was active until 1970,and was the birthplace of companies such as Antar and Schlumberger Limited. So far.

While the Asian continent around 900 B.C., gas use is *known in China*. It is precisely in this country that the first known natural gas well of more than 100 meters deep is made, through a rudimentary system of bamboo reeds and percussion bits. There is evidence that they burned the gas to dry the salt rocks they found between the limestone layers. In the 7th century in Japan the existence of a gas well was also discovered.

In Europe, gas was not known until its discovery in England in 1659, although its use was not mastified. The first use of natural gas in North America was made from a well in New York State in 1821, it was a shallow well and the gas was distributed to consumers through a small-diameter lead pipe, for cooking and lighting. During the 19th century, the use of natural gas remained localized because there was no way to transport large amounts of gas over long distances, which is why natural gas remained displaced from industrial development by coal and oil. With the advancement in the design of pipes it is feasibility to make it feasibl to transport GN over long distances during the first half of the twentieth century. After World War II, better pipes and pipelines of greater diameter and length are developed and in the 1970s, the largest pipeline is built from Russia to Eastern Europe with almost 6,000 km in length.

Around 1920 is when the first techniques of liquefaction of air gases are developed the great revolution for the transport and marketing of GN occurs. Initially the GN was liquefied for the production of the associated helium in its composition and was later regasified for introduction into the pipeline and its distribution, with the development of cryogenic technologies a new world was opened for transport (1 m3 of LNG equals 580 m3 of GN) for its high concentration of volume in liquid state.

3 EXPLORATION AND EXPLOITATION

The existence of a concentration of ore, element or rock with sufficient economic value to sustain the holding must comply with a process of predictability

After a deposit has been discovered, explored, outlined and evaluated, the next step will be the selection of the mining method that physically, economically and environmentally adapts for the recovery of commercially valuable ore. From an economic point of view, the best method of exploitation should be one that provides the highest rate of return on investment.

To ensure this return the traits and characteristics of mineral deposits play a fundamental role. This will depend on the conditions that determine the most appropriate mining method. From the point of view of structural geological engineering, the following characteristics are of paramount importance in the selection of a mining method:

- The size and morphology of the mineral body.
- The thickness and type of surface escarp.
- The location, course and blight of the reservoir.
- The physical characteristics and strength of the mineral.
- The physical characteristics and strength of the rock in the winding.
- The presence or absence of groundwater and its hydraulic conditions related to the drainage of the works.
- Economic factors involved with the operation, including the law and type of ore, comparative mining costs and desired production rates.
- Ecological and environmental factors such as conservation of the original geomorphological contour in the mining area and prevention of harmful substances contaminating the waters or atmosphere.

In addition, the selected method must meet maximum safety conditions and allow an optimal extraction rate under the particular geological conditions of the tank. Mining methods should be developed on the basis of structural geology and rock mechanics, with the fundamental concept of stability prevailing in the works.

In recent times in operating contracts methods have been added for a recovery of post-exploitation areas, we know this as mine closure, where environmental liabilities must have a visually aesthetic and innovative ecological final treatment, however one of the main drawbacks are the long operating periods and the poor control of government agencies to monitor the sustainable demands of the economically profitable area.

EXPLORATION TOOLS AND TECHNIQUES

Mining and oil exploration and exploitation is based on a number of techniques, instrumental and other empirical, at a very different cost. Therefore, they are usually applied successively, only if the value of the product is sufficient to justify its use, and only if they are necessary to complement the techniques that have already been used so far.
Gathering information

It is one of the preliminary, low-cost techniques that can be carried out in the office itself, although in some cases it involves certain displacements, to locate the information in purely documentary external sources. It consists basically of collecting all the information available on the type of site prospected (geological characteristics, volumes of expected reserves, geometric characteristics), as well as on the geology of the study area and its mining history (type of mining that have existed, volume of productions, causes of the closure of the holdings). All this information should allow us to establish the specific model of deposit to be prospected and the conditions under which the prospecting process should be carried out.

At this stage it is very useful to have the support of metalgenetic maps that show not only the location (and typology) of deposits, but also the relationships between them and their environment. In this sense, it is very useful to graphically represent metal parts or metalgenetic provinces in these.

Remote sensing

The use of information from artificial satellites orbiting our planet can be of great interest in mining research. It remains a relatively low-cost technique (conditioned by the price of the information to be collected from the agencies that control this type of information) and that is applied from the cabinet, although also often supplemented by field outings.

The information provided by satellites that is of geological-mining utility refers to the reflectivity of the terrain versus solar radiation: it affects the terrain, is partly absorbed, and partly reflected, depending on the characteristics of the terrain. Certain radiations produce the sensations appreciable by the human eye, but there are other areas of the electromagnetic spectrum, invaluable to the eye, that can be collected and analyzed by specific sensors. Remote sensing takes advantage of precisely these spectrum bands to identify characteristics of the terrain that may reflect data of mining interest, such as alterations, presence of certain minerals, temperature variations, humidity, among others.

Geology

The study in greater or lesser detail of the characteristics of a region is always necessary in any study of the mining field, since each type of deposit usually presents specific conditions that must be known in order to carry out with greater guarantees of success our exploration, as well as others that can be undertaken in the future. It is a study that is carried out during the pre-exploration and exploration phases, as its cost is still quite low. It also has a dual aspect, in the sense that it can partly be done in cabinet, based on information collection and remote sensing data, but when it needs some detail, it must be supplemented with observations on the ground.

The generic term geology encompasses many distinct sections of the geological recognition work of an area. Geological mapping (or elaboration of a geological map of it) includes stratigraphic surveying (knowing the succession of stratigraphic materials present in the area), tectonic study (identification of tectonic structures, such as faults,folds, affecting materials in the area), petrological study (correct identification of different types of rocks), hydrogeological (identification of aquifers and their most relevant characteristics), etc. In each case they will be of greater or less importance to each other, depending on the specific control presented by the investigated mineralization.

Geochemistry

Geochemical prospecting consists of the analysis of sediment samples from streams or soils or water, or even plants that can concentrate chemical elements related to a given mineralization. It has its basis that the chemical elements that make up the bark have a characteristic general distribution, which, although it may be different for each different area, is characterized by presenting a range of values defined by a log-normal unimodal distribution, in other words, the "normal" concentration of that element in samples in a region appears as a gaussian bell in a semi-lizard graphic. However, where there is some abnormal concentration of a given element in the area (which may be produced by the presence of a mineral deposit of that element), this distribution is altered, usually giving rise to a bimodal distribution, which allows to differentiate the normal populations (the existing one in the mineralization environment) and anomalous (which will be placed precisely on mineralization).

Thus, the different variants of this technique (soil geochemistry, streams, biogeochemistry) analyze samples from each of these environments, following ordered patterns, so that it is possible to have a representative analysis of an entire region, in order to identify the anomalous population(s) that may exist in it, and differentiate them from possible abnormal populations that may be an indication of the existence of mineralizations.

The cost of these techniques is usually higher than those of a geological nature, since they involve a team of several people for the taking and preparation of samples, and the cost of the corresponding analyses. Therefore, they are applied when geology already provides information that makes it possible to fundamentally suspect the presence of deposits, and confirms or discards a specific type of anomaly.

Geophysics

Within this generic name we find, associated with geology, a whole range of very diverse techniques, both in cost and in applicability to each specific case. The base is always the same: try to locate rocks or minerals that present a physical property that contrasts with that of the minerals or encompassing rocks. Just as locating a needle in a haystack a magnet is a useful tool, this same magnet will be of no use to us if what we have lost between the straw is a 0.5 mm pen mine.

Thus, the various applicable techniques and their field of application can be as follows:

Electrical methods: They are based on the study of the conductivity (or inverse, resistivity) of the terrain, using relatively simple devices: a system of introduction of current to the ground, and another of measurement of resistivity/conductivity. They are used to identify materials of different conductivity: for example, sulphides are often very conductive, just like graphite. They are also widely used for water research, because rocks containing water become somewhat more conductive than those that do not contain water, as long as the water has a certain salinity that makes it in turn conductive.

Electromagnetic methods: It is based on the study of other electrical or electromagnetic properties of the terrain. The most commonly used is the induced Polarization method, which involves mediating the effect of ground overload: a high voltage electric current is introduced into the ground and when it is interrupted it is studied how the ground is loaded, and how the electric shock process occurs. Widely used for sulfide prospecting, as they are the ones with the highest loads. Other techniques: spontaneous polarization, magnetoteluric methods, etc.

Magnetic methods: Based on the measurement of the magnetic field in the field. This magnetic field as we know is a function of the Earth's magnetic field, but can be affected by existing rocks at a given point, especially if they exist in the same ferromagnetic minerals, such as magnetite or pyrrotin. These minerals cause an alteration of the local magnetic field that is detectable by so-called magnetometers.

Gravimetric methods: are based on the measurement of the earth's gravitational field, which, as in the previous case, may be modified from their normal values by the presence of specific rocks, in this case of density other than normal. The gauge is the instrument used to detect these variations, which because of its small entity and the influence of topographic variations require very detailed corrections and therefore also very expensive. This technique has been used with great effectiveness in the detection of mass sulfide bodies in the Iberian Pyritic Sash.

Radiometric methods: are based on the detection of radioactivity emitted by the terrain, and are mainly used for the prospecting of uranium deposits, although exceptionally they can be used as an indirect method for other elements or rocks. This field-emitted radioactivity can be measured either on the ground itself, or from the air, from planes or helicopters. The most common measuring instruments are basically of two types: Scintilometers (also called twinkle counters) or Geiger counters. However, these instruments only measure total radioactivity, without discriminating against the wavelength of the radiation emitted. The most useful are sensors capable of discriminating against different wavelengths, because these are characteristics of each element, which allows to discriminate the element that causes radioactivity.

Seismic: The transmission of seismic waves through the terrain is subject to a series of postulates involving parameters related to the nature of the rocks they pass through. In this way, if we cause small seismic movements, by explosions or fall of heavy objects and analyze the distribution of seismic waves to strategically placed measurement points, as is done with sound waves in ultrasounds, we can draw conclusions about the nature of the rocks of the subsoil. Two different great techniques differ: reflection seismic and refractive, which analyze each of these aspects of seismic wave transmission. It is one of the most expensive techniques, so it is only used for researching high-cost resources, such as oil.

Geology has a whole range of different tools that are very useful, but that must be applied to each specific case according to two parameters: its cost, which must be proportional to the value of the object of exploration, and the technical feasibility, which must be considered in the light of the preliminary analysis of the physical characteristics of this same object.

Calicatas

Often, after the application of the above techniques we still have reasoned doubts as to whether or not what we are investigating is something with a mining interest. For example, we may have a lead geochemical anomaly and an electrical geophysics anomaly, but will it be a galena mineralization or an old buried pipe? In these cases, to verify at low cost our interpretations of alignments of possible mining interest can be made trenching in the field by means of backhoe shovel, which allow to visualize the rocks located just below the analyzed or recognized ground. In addition, these calicatas will allow to obtain more representative samples of what exists in the subsoil, although it should not be forgotten that because of their small depth of work (1-3 meters, at most) they are still not comparable to what may exist below the level of metheoric alteration, since, as we saw in the corresponding section, precisely mineralizations tend to favor supergenic alteration.

Mechanical surveys

Surveys are a vital tool mining research, which allows us to confirm or deny our interpretations, since this technique allows to obtain samples of the subsurface at variable depths. Its main problem stems from its representativeness, since it should not be forgotten that these samples constitute, at best (probes with continuous core recovery) a rock cylinder a few centimeters in diameter, which may not have fully recovered (there may have been losses during drilling or extraction), and that it may have cut the mineralization at an exceptionally poor or exceptionally rich point. However, they are the most valuable information available on mineralization as long as it is not reached through mining work.

Mechanical surveys are a very complex world, in which there is a wide range of possibilities, both in terms of the drilling method (percussion, rotation, rotopercussion), and in terms of working diameter (from metric to millimeter diameters), in terms of the range of achievable depths (which can be thousands of meters in oil probes), as for the extraction system of the cut material (continuous witness recovery, drag through the drilling water, or by compressed air). All this makes conducting mechanical surveys a particularly important stage in the mining research process, and requires more detailed and problematic decision-making.

After taking the data, in any of the available methods, the mining exploration process must be interpreted, so that every decision made whether or not to follow the next steps is based on data that supports our preliminary interpretation or not.

In this way, each stage of the research we develop must be aimed precisely at supporting or denying preliminary interpretations, through new data that imply an improvement in interpretation, but without systematically seeking confirmation at all costs of our idea, sometimes misinterpretations, bad practices or very subjective positions can be very costly for the company, although without it there would often be no mining research, remembering that most geologists are antagonists of nature.

In short, the interpretation of the results should be very detailed, and it must look for the coincidences that support our ideas, but also the non-coincidences, which must be analyzed in a particularly careful way, looking for the alternative explanation(s) that may involve the confirmation or dementedness of our interpretations, without forgetting that in the end the polls will confirm them almost definitively.

Four basic principles of mining or oil exploitation are currently **identified**:

1. Surface or open sky

Used for the extraction of metallic and non-metallic minerals from mineral bodies located at depths less than 160m (500 feet approx.)

- Minado de placers. Concentration of heavy minerals from detritic materials.
    - ✓ Troughs and gutters
    - ✓ Hydraulic mining
    - ✓ Dredging
- Mined to Open Tagus (open sky). Any type of ore deposit in any type of rock, located on or near the surface of the terrain.
    - ✓ Individual bank
    - ✓ Multiple banks
    - ✓ Cloak stripping
    - ✓ Quarrying
- Glory Hole. Open pit excavation from which the ore is removed by gravity through one or more counterporas at underground haul levels.

2. Underground Mining.

Exploitation of mining resources that develops below the surface of the land. For the selection of this method several factors should be considered such as resistance of the mineral and the rock boxing, size, shape, depth, buzzling angle and position of the tank; continuity of mineralization, among other specific geological characteristics in the area of interest.

- Naturally supported cuttings. Excavations in which the loads exerted by the rock on the opening are supported by the carved walls or pillars of the same rock.

    ✓ Open rebates
    ✓ Halls and pillars
    ✓ Tumbe by sublevels
    ✓ Lie down on load
    ✓ Open recesses with horizontal locks

- Artificially supported downs. Work in which a significant part of the load or weight of the surrounding rock is supported by some artificial support (struts, frames, fillings, etc.).

    ✓ Cut and fill
    ✓ Conjugate tables
    ✓ Long fronts
    ✓ Short fronts
    ✓ Descending slices

- Sinking slices. Applicable to massive mineral deposits with large horizontal developments likely to collapse to follow the sinking of the ore as it is removed and extracted.

    ✓ Sub-level subsidence
    ✓ Sinking of blocks and panels

3. Special Methods

- Indirect methods are a series of systems that employ techniques for dissolving the values contained in the deposit, it is not necessary to physically penetrate the deposit for extraction.

    ✓ Frasch Process

- ✓ Dissolving with hot water
- ✓ Leaching

It will be important to consider in the decision to operate a mine by underground or surface methods the activities of sweeping, blasting, laden and transporting rocky material subject to the holding, including crushing the ore. Losses in mineral recovery should also be taken into account as they are greater in underground mining than in the surface, affecting the productive life of a mine.

4. Drilling Wells

Today rotary drilling is the most widely used in oil **wells.** This method uses cylindrical steel tubes coupled to a rotating drum or table, by which they are printed a rapid rotation, mixed with weight and pressure of the drilling fluids achieve a breakthrough at great depths, in some cases operational difficulties become present, and different specialized techniques must be applied to drilling.

To achieve a breakthrough in drilling there must be prior study of pressures of formations, geological zones, type of rocks, environments, nodal analysis, among other things.

A well, when it has been drilled and piped to the area with oil, is ready to start producing. If the natural pressure of the gas is high, the oil is quickly driven from the bottom and goes up the pipe.

After the drilling process occurs the completement system, and a series of tools to move the gas or oil from its original site to the surface, this movement requires an analysis, for all the loss of pressures of the formation or friction that is generated internally by the different materials.

In order to regulate, without loss, the flow of oil through the mouths of wells, a valve system called the 'Christmas tree' has been created. However, in many deposits, additional measures must be taken to ensure that the well is put on or kept in production, lowering to the bottom a production pipe of relatively small diameter (between five and ten centimeters) to control the oil or gas outlet.

When well pressure is not sufficient for oil to rise to the surface, artificial production and lifting systems are also used. Among these, the most common is mechanical pumping, easily recognizable on the surface by the presence of the pumping unit. There are also other pumping systems, such as electrocentrifuge, pneumatic (gas lift) and hydraulic.

Groundwater usually causes major challenges in the oil industry; however, they also contribute to expelling oil to the surface, but it is common for the pressure to push oil into the well to gradually decrease and production to the point where the well produces no more, leaving appreciable amounts of oil in the underground without recovering. In these cases, it helps to recover more oil by injecting gas, water or other fluids into the reservoir.

Oil is transported from the well, through pipes, and special tools, in which gas and water are separated. From separators, pipes (gas pipelines) drive the fluid to different sites for use as fuel or for further treatment. Another pipe (pipelines) drives the fluid to the storage tanks from where it will be sent to its destination, either a refinery or a port of embarkation.

One of the major deployments of technology development is oil refineries, the separation process takes place in fractionation or primary distillation towers. There are several processes for the treatment of hydrocarbon, all use heat as a means of separation when heating it. Thus, as the temperature rises, compounds with fewer carbon atoms in their molecules (and which are gaseous) detach easily; then the liquid compounds are vaporized and also separated, and so on, the different fractions are obtained.

To do this, the crude oil is first heated to 400oC so that it enters the distillation tower. Here the vapours rise through floors or compartments that prevent liquids from passing from one level to another. As the floors rise, the fumes cool down.

This cooling results in different fractions condensing on each floor, each of which has a specific liquefaction temperature.

The first vapors to be liquefied are those of the heavy diesel at approximately 300oC, then the light diesel at 200oC; then kerosene at 175oC, naphtha and finally gasoline and combustible gases coming out of the fractionation tower still in the form of steam at 100oC. This last fraction is sent to another distillation tower where gases are separated from gasoline.
However, in this fractionation tower it is distilled at atmospheric pressure, that is, without pressure. Therefore, hydrocarbons containing 1 to 20 carbon atoms can only be separated without decomposing.

In order to recover more fuels from the residues of the primary distillation it is necessary to pass them through another fractionation tower that works at high vacuum, that is, at pressures below that of atmospheric to avoid their thermal decomposition, since the hydrocarbons will be distilled at a lower temperature.

Only two fractions, one of distillates and one of debris, are obtained in the vacuum tower.
Depending on the type of crude oil being processed, the first fraction contains the hydrocarbons that make up lubricating oils and paraffins, and the residues are those with asphalt and heavy combustoleum.

Table 1 describes approximately the number of carbon atoms containing the different fractions mentioned above.

fracción	núm. de átomos de C por molécula
gas incondensable	$C_1 - C_2$
gas licuado (LP)	$C_3 - C_4$
gasolina	$C_5 - C_9$
kerosina	$C_{10} - C_{14}$
gasóleo	$C_{15} - C_{23}$
lubricantes y parafinas	$C_{20} - C_{35}$
combustóleo pesado	$C_{25} - C_{35}$
asfaltos	$> C_{39}$

Table 1. Mixture of hydrocarbons obtained from the fractional distillation of petroleum

Of the incondensible gases methane is the lightest hydrocarbon, as it contains only one carbon atom and four hydrogen atoms. The next one is ethanol, which is made up of two carbon and six hydrogen.

The first is the main component of natural gas. It is usually sold as fuel in cities, where there is a network of special pipes for distribution. This fuel contains significant amounts of ethanol.

LP gas is the fuel distributed in stationary cylinders and tanks for homes and buildings. This gas consists of hydrocarbons of three and four carbon atoms called propane and butane respectively.
The next fraction consists of virgin gasoline, which consists of hydrocarbons of four to nine carbon atoms, most of whose molecules are distributed linearly, while others form cycles of five and six carbon

atoms. This type of compound is called paraphanics and paraphanic cycles respectively.

The fraction containing 10 to 14 carbon atoms has a boiling temperature of 174 to 288oC, which corresponds to the kerosene-called fraction, from which fuel is extracted from turbine aircraft called turbosin.

The last distilled fraction of the primary tower is diesel, which has a boiling range of 250 to 310oC and contains 15 to 18 carbon atoms. From here you get the fuel called Diesel, which, as we said, is suitable for vehicles that use Diesel engines such as tractors, locomotives, trucks, trailers and boats.

Of the vacuum distillates obtained, those that by their characteristics are not intended for lubricants will be used as raw material to convert them into light fuels such as liquefied gas, high octane gasoline, Diesel, kerosene and diesel.

Vacuum residue contains the fraction of heavy fuels used in thermoelectric boilers.

Almost every barrel of oil processed at refineries goes to fuel manufacturing. The amount of virgin gasoline obtained depends on the type of crude oil (heavy or light), since in each case the percentage of this fraction is variable.

As we said at the beginning, gasoline is the fuel that is most in demand; therefore, the amount of natural gasoline obtained from each barrel is always insufficient, even if light crude oils are destiled, which have up to 30% of this product. In addition, the characteristics of this gasoline do not fill the octane specifications required for automobile engines.
To solve these problems scientists have developed a series of processes to produce more and better gasolines from other fractions of oil.

But these solutions must be accompanied by a *green solution,* it is true that we must seek to make efficient the consumption of the final product, which will help us to assess and delay the exploitation of natural resources, therefore, we will have more products for longer, consumption-life ratio.

These solutions must be accompanied by proposals aimed at environmental recovery and improvement of the quality of life of most societies, which leads us with an in-depth analysis of our usefulness and syncretism on the part of the countries of the hemisphere, only by applying the laws that are already in place, is only to apply and control them by separating the corrupt culture that engapples us.

If we achieve sustainable awareness, we will secure the home of future generations, leave them the Millennials challenge.

4 THE ROLE OF EXPLORATION AND EXPLOITATION RESEARCH IN INDUSTRIES; MINING AND OIL

Mining and oil industries are paramount activities for the growth of our societies. But we also have sectors that are eager to develop green technology such as metalworking, agriculture, livestock, fisheries, computing and communications, among others, it demonstrates the importance of this sector for the growth of our countries.

Clearly, the world has evolved and that standards in every order of things have also evolved. There are no doubt that environmental and social responsibility standards impose challenges that we have and want to face

Scientific and technological research, from an economic perspective, is appreciated to the extent of its contribution to the value of products and services. In this logic, knowledge is incorporated into production processes through several operations:

1. **Technology transfer**
2. **Production systems**
3. **Marketing**
4. **Marketing**
5. **Business management.**

Meanwhile, the most developed countries establish their competitiveness by articulating the knowledge-generating system, the production system and services.

Thearticulation gives rise **to innovation systems,** as well as a set of social and economic relations summarized in the knowledge *society*. This is how they become exporters of raw materials and importers of knowledge, and we become producers of non-renewable matter and consumers of knowledge, but with limited development.

4.1 MINERIA

DEVELOPMENT OF RESEARCH AND APPLICATION OF INNOVATION IN MINING

For years dozens of researches have been conducted on advances for productive improvement. We have also highlighted the great benefits it brings to society, but also by studying its resilience on the part of communities and the population.

It is incomprehensible that our region will provide a large percentage of global production and large economically profitable geological deposits and lack cooperation for technology development for our greatest benefit, even in the new millennium the parceling of benefits remains, and worse the lack of concern to resolve environmental liabilities.

According to ECLAC (Economic Commission for Latin America and the Caribbean) the region has significant feasible reserves (Graphic 1). Where we can indicate one of the resources we have. For example, copper reserves are estimated to amount to 2.3 billion tonnes on land and marine land. Data confirmed by geological survey of the US geological survey indicate 40% of these tonnes in Chile's possession, also 30% of Bauxite reserves, 41% nickel and 29% of silver reserves. It was estimated that these reserves could be much larger.

Peru is the country with the greatest mining potential in the region, in terms of projection, followed by Brazil, Mexico, Argentina, Venezuela and Bolivia, these countries are at the geological target of the region attracting greater investment and international interests. This should result in maximizing the development and use of technology to estimate our potential, which is in proven reserves. They should translate them into sustainable benefits, and this is only possible if we apply futuristic environmental conservation. The stimulation to carry out research, development, application and innovation lies under our feet.

Figure 1. Latin America and the Caribbean, a significant share of global reserves of major minerals Metal

Source: Economic Commission for Latin America and the Caribbean (ECLAC), based on USGS Mineral commodity summaries 2018

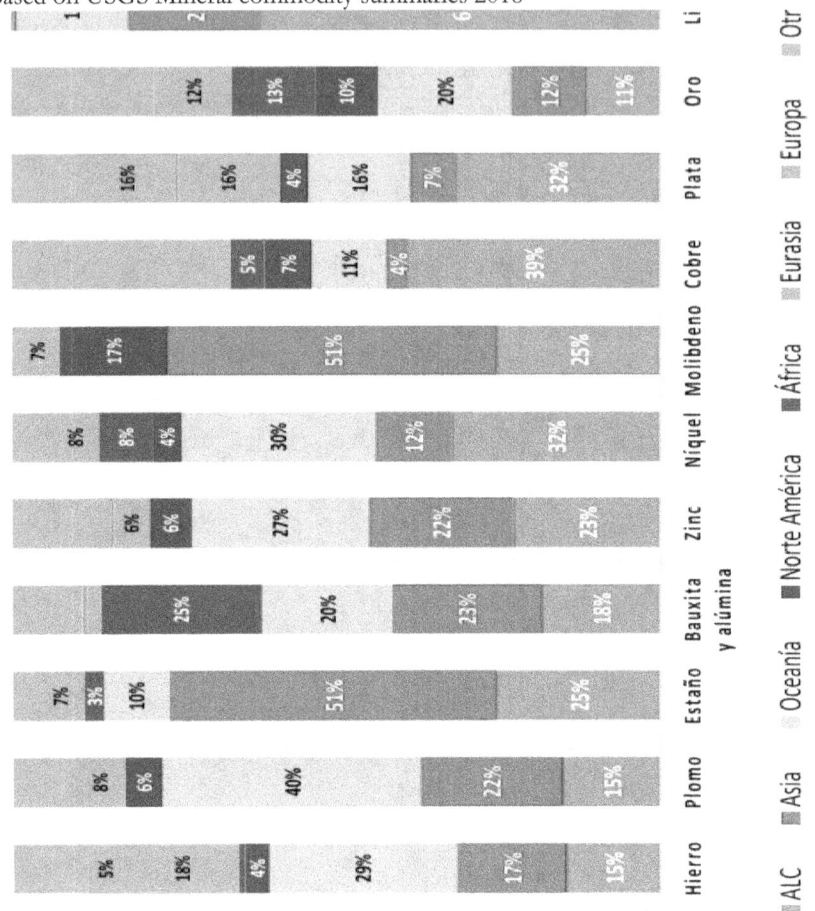

Growth rates from the 1990s to the present day have been above the global production growth rate and in many cases, we are among the 3 most producing countries on the planet. (Table 2)

Table 2: Participation of Latin America's major mining countries in the production.

Pais	Participación en la producción mundial Año 2004
Argentina	12° productor de cobre mina
	14° productor de estaño refinado
	15° productor de plata mina
Bolivia	4° productor de estaño mina
	6° productor de estaño refinado
	11° productor de plata
	13° productor de zinc mina
Brasil	1er productor de hierro
	2° productor de bauxita
	6° productor de aluminio primario
	5° productor de estaño mina
	7° productor de estaño refinado
	10° productor de níquel mina
	14° productor de níquel refinado
	13° productor de oro
	14° productor de zinc mina
	14° productor de zinc refinado
Colombia	8° productor de níquel mina
	8° productor de níquel refinado
Cuba	6° productor de níquel mina
	10° productor de níquel refinado
Chile	1er productor de cobre mina
	1er productor de cobre refinado
	6° productor de plata
	15° productor de oro
Guyana	12° productor de bauxita
Jamaica	3° productor de bauxita
México	11° productor de cobre mina
	14° productor de cobre refinado
	12° productor de hierro
	1er productor de plata
	5° productor de plomo mina
	6° productor de plomo refinado
	7° productor de zinc mina
	8° productor de zinc refinado
Perú	3er productor de cobre mina
	9° productor de cobre refinado
	3° productor de estaño mina
	3° productor de estaño refinado
	7° productor de oro
	2° productor de plata
	4° productor de plomo mina
	12° productor de plomo refinado
	3er productor de zinc mina
República Dominicana	11° productor de níquel mina
	13° productor de níquel refinado
Suriname	10° productor de bauxita
Venezuela	8° productor de bauxita
	12° productor de aluminio primario
	10° productor de hierro
	14° productor de níquel mina

Source: World Bureau of metal Statistics, Banco Mundial y CEPAL.

These numbers give us a privileged position among the best international competitiveness in development, technology and knowledge transfer on the planet, and our efforts lead us to understand the relationship of profit in the image of our countries if we take exploitation seriously with environmental responsibility.

For this we have some challenges to overcome:

- Informal and illegal high-risk activities (mercury)
- Contamination of water, air and soil from extraction, smelting and transport processes.
- Competition for water use (watersheds and reservoirs)
- Habitat destruction and protected areas
- Overlapping mining areas over areas of importance to biodiversity
- Numerous environmental liabilities
- Surface runoff, infiltration and acid drainage.
- Dragging particulate matter.
- Extreme events and physical stability of the relay deposits.

These conflicts definitely damage the image to our customers, a sign of how bad we are we can see them on Map 1, where environmental commitments are still present in the region and the poverty of waste management (Map 2) remains outstanding.

A new environmental planning model

Map 1: Removing minerals and building materials

Source: Economic Commission for Latin America and the Caribbean (ECLAC), on the basis of Environmental Justice Atlas.

Map 2: Waste management.

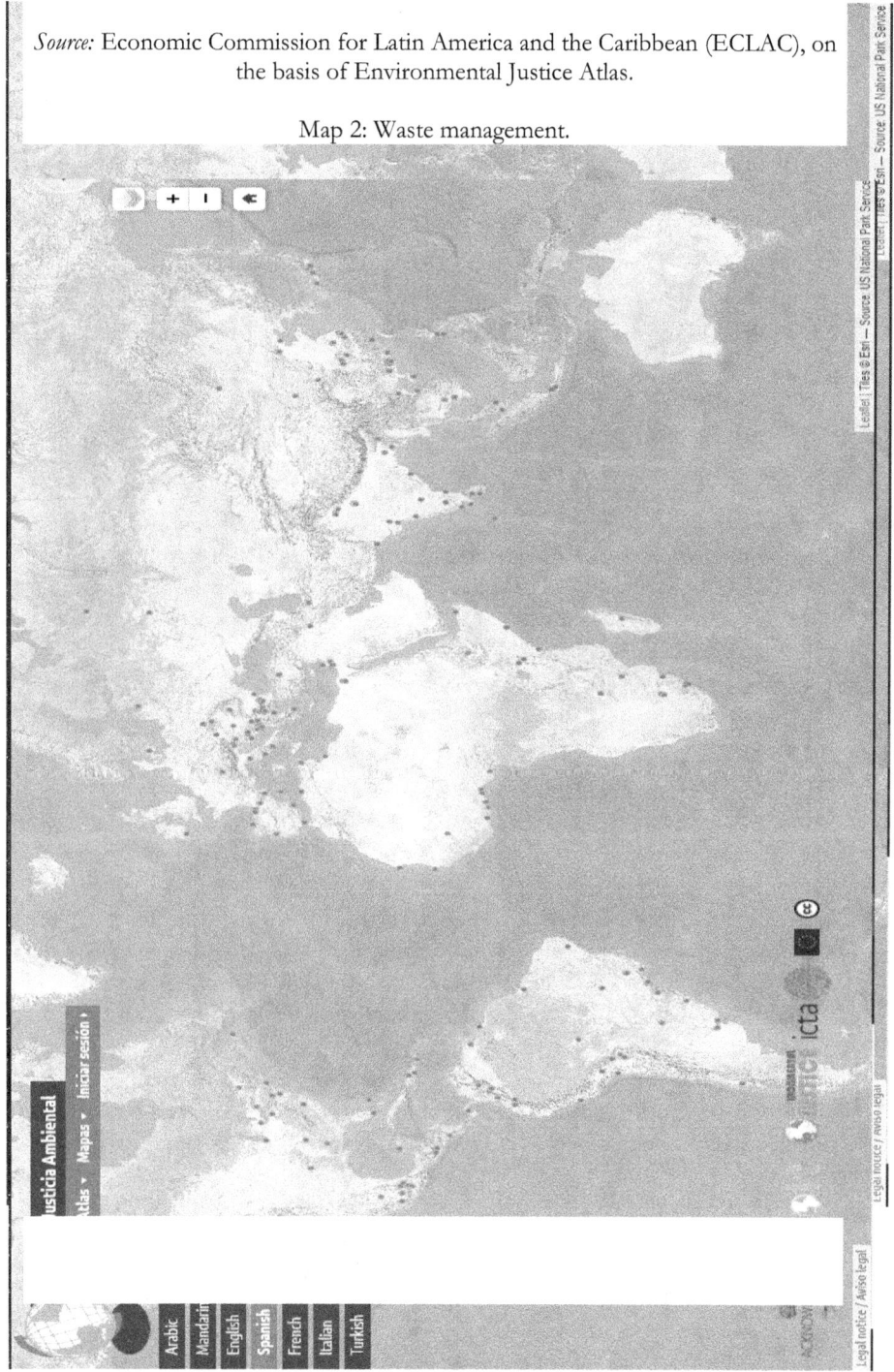

Source: Economic Commission for Latin America and the Caribbean (ECLAC), on the basis of Environmental Justice Atlas.

We have a duty to sow a culture of research in professionals in the different productive sectors, which must come from the classrooms of internalized classes as part of the ethical and moral chairs, linking the avoidable cooperation of productivity – security – social/environmental responsibility.

All of the above, which we can take as an example, should lead us to look for alternatives that will reduce this environmental impact.

Breaking work paradigms can be done with the use of new technologies, energy sources, concepts in each of the stages of operation and production processes. Similarly, it creates new solutions or novel outlets that can strengthen the industry at large.

That's how we aim for it. Research and development also allow new products and services to be offered to solve new problems. But not only that, in its most essential form, scientific research is a tool that allows us to use our intelligence to solve problems by adding value to productivity.

4.2 PETROLEO

INNOVATION AND TECHNOLOGY NEEDS FOR THE OIL AND GAS INDUSTRY

The oil and gas industry are undergoing a shift towards new forms of hydrocarbon exploration and production in complex geological environments and socio-environmentally sensitive areas. For its part, the world is demanding more and more energy to sustain its growth, so the industry faces the dual challenge of reinventing itself to meet demand and make it responsible in the economic, social and environmental dimensions.

Operations are increasingly socially impacted and are developing at a very filimitnot between wealth generationandthe side effects of migration to previouslydepopulated areas and with local economies that are affected by oil activities.

It is necessary to concert the visions of our region in the global context of the oil and gas sector, as well as the main needs and challenges where innovation and technology, in a collaborative environment between academia and national industry, can contribute, where it seems that these two are in a divorce in time and space.

This dichotomy represents a challenge to the South American academic sector to include pragmatic training activities that prepare professionals not only with technical excellence but with better capacities of relationship, environmental awareness and ethical behavior, with knowledge of local opportunities and with greater entrepreneurship, that allow to fill the gap of small companies based technologically that in other countries contribute significantly to the creation of new products and technologies that support the sector.

This is where cooperation between the countries with the highest production can take advantage of exporting their geniuses and expanding the proportions of jobs by making opportunities equitable and equal, so South America would become an exporter of knowledge among the main oil-producing countries of the region, Venezuela, Brazil, Mexico, Colombia, Ecuador, and Argentina. The demands of the industry and the accelerated pace of the planet can never be removed from classrooms, that growth must involve teachers, researchers, students, representatives and managers.

Reviewing our curriculum should go at the same speed of technology creation and innovation development. The need for intensive capital, the high level of risk and the design of field testing could explain thissituation, therefore it is necessaryto close gaps and shorten cycles so that technological solutions arrive in a timelier andreliableway, especially in the fields ofproductionorn.

GREAT CHALLENGES OF THE OIL AND GAS INDUSTRY

In the face of depletion of oils found in easily accessible deposits, the available resources are concentrated in heavy, extra-heavy hydrocarbons and unconventional deposits, where large volumes are calculated, but with technologicaland economicdisaffiphysitivesfortheir production (Labastie, After & Holditch, 2009). The industry must reinvent itself in order to meet the expected demand.

Innovation and technology have a major impact on industry and society, making large companies quickly created or disappearing because of their effects. Every technological development achieved, as well as those planned for the coming years (Manylka, Chui, Bughin, Dobbs, Bisson & Marrs, 2013), have a common denominator, the need for energy, and for now it is clear that fossil hydrocarbons such as coal, oil and gas will remain, at least for the next 30 years, the main source to meet the demand for energy and mobility on the planet.

According to the U.S Energy Information Administration (EIA), global energy consumption will increase by 56% between 2010 and 2040, with fossil hydrocarbons being the main source with a close share of 80% (International Energy Outlook – IEO, 2013).

Also noteworthy is the projection of the increase in oil production from 87 to 115 million barrels per day in 2040, mainly for use in the transport and industry sectors, as well as the growth in natural gas consumption, the additional production of which would come from the development of unconventional deposits (tight gas, shale gas and methane associated with coal mantle).

Everything indicates that hydrocarbons will continue to be the main source of energy on our planet. In terms of hydrocarbon supply, the most notable phenomenon remains the U.S. shale oil and gas revolution. In 2012, the U.S. recorded the largest increase in oil and natural gas production in the world, the highest figure in oil production in its history (British Petroleum – BP, 2013).

For its part, given its position and melting conditions as a result of climate change, Russia has declared its interest in exploring hydrocarbons in the Arctic, which by their magnitude have been changing the global geopolitical landscape and fueling potentialswith international flictos (Andres, 2010). According to the U.S. Energy InformationAdministration (EIA, 2014), about 22%of theworld'shydrocarbon reserves are in the Area, some 412 billion barrels of oil equivalent, of which78% would be natural gas. For the U.S. Geological Survey, the Russian Arctic continental shelf contains more than 20% of the world's uncovered crude oil and natural gas resources.

This is one of the great examples where technological innovation in the oil and gas industry supports the sector with the necessary equipment and practices to continuously increase production and provide the petrochemical fuels and by-products that society demands, even in the most inhospit environments.

The oil and gas industry has advanced technologically to improve its processes. In the 1950s the industry focused on technologies to find and produce oil in areas with significant surface oil manifestations, processing the lightest possible crudes in refineries whose operation was very manual. In the 1980s with the rise of 3D seismicism, exploration emerged in more complex areas of geology, the development of horizontal wells and the support of information technologies; In addition, refinery diets and products were diversified, including medium crude oil processing, increased diesel production, and the environmental and process safety approach was strengthened. Today, technological challenges focus on leveraging unconventional resources, increasing the recovery of existing fields, and using increasingly powerful and portable resources from mass computing and wireless technologies. All of these with a strong focus on process safety, environmental management and fuel quality.

On the environmental issue, today humanity uses the equivalent of 1.5 planets to provide the resources it uses and absorb waste, meaning that it takes one year and six months for Earth to regenerate what is used in a year. Moderate UN scenarios suggest that, if current trends demográficas and consumption continue, by 2030 wewillneedthe equivalent of two Earth planets, and of course, thereisonlyone (Global Footprint Network, 2014). Environmental challenges remain a great source of research and technological development if all businesses, particularly oil and gas, are to be made sustainable.

TECHNOLOGY IN UPSTREAM, MIDSTREAM AND DOWNSTREAM

What would be the projection for closing technological gaps and aligning with today's global challenges. The main technological challenges throughout the value chain of the Oil & Gas business, in which the academy and industry can contribute:

In the upstream (exploration and production): reduction of geological risk and improvement of the image of the subsoil, increase of the recovery factorand optimization costs of production, managementand ficient of water, and proof of the potential of unconventional deposits.

*In the midstream (transport*and*)* evacuation of heavycrude oils, assurance of the withtcapacity andintegrityofthe infrastructure, andconsolidation in the biofuels market.

In the downstream (refitionandmarketing): valuationof heavy crude oils, improvementoffuelquality, increase in diesel production, and decrease in fuel oil performance. For Information Technology (IT), information issearched withtable and secure in real time through dataavailable daily and in an automated manner. In the face of these challenges, key technologiesmustbedesignedto strengthen, incorporate and ensure longevity, depending on the demographic characteristics of each country. We can glimpse some.

3D seismic and watershed modeling– reduce exploratory uncertainty, optimize development wells and define areas of potential hydrocarbon production. Basin modeling, on the other hand, allows for a better prediction of rock characteristics and the type of fluids in the subsurface to decide which basins to invest in with the least risk.

Improved recovery *and* technology for optimization*or development*costs: its impact is to make heavy crude oil production feasibling, increase therecovery factor and reduce drilling costs by at least.

Technologies for water management (bottom, surface and environmental control): its impact is to reduce water in superficie between 15% and 20% and makeit possible tomonetizeits valuationas a resource. Its application would reduce the pouring and increase re-injection for recovery.

Contact maximisation technologies: they would make unconventional reservoirs feasibling.

Dilution: Viscosity and friction reducers. The impact is caused by the reduction in naphtha consumption with the respective savings in the transport of thinner by the ducts.

Infrastructure risk management technologies: technologies that mitigate risks associated with threats of ground movement failures and external corrosion.

Energy cogeneration and biofuel production: these technologiesensureenergysupply and science and consolidatethe region in the biofuels market.

Crude oil improvement: It would be able to increase the APIseverity of crudes from 8 to 20 and decrease naphtha imports.

Sulphur reduction: Fuel qualitycanbe counted under Euro 4 and Euro 5 standards.

The current capacity for these technologies is medium-low relative to world leaders (high-medium) and their incorporation must be given through a combination of different mechanisms, including the purchase of technological services, adaptation, research and development.

GREENHOUSE GAS MITIGATION

One of the great liabilities we need to solve is the management of greenhouse gas (GHG) emissions, it has become an issue of great interest in recent decades, due to the operational, economic, environmental and regulatory burdens that impact and affect reputation and relationships with business stakeholders. In addition, there are sectoral, international policies around clean production, non-carbonization of the economy and carbon taxes. For this reason, companies around the world have begun investing in actions aimed at monitoring, controlling and/or reducing GHG emissions (direct and indirect), seeking to identify mitigation opportunities in the full value chain of their businesses, as well as for those associated with the use of their products. During 2012, power generation plants were located in the sector with the highest GHG emissions, followed by the oil and gas industry.

It is vitally important that large companies in the region (PDVSA, Ecopetrol, Petrobras, Petroperú, PECU, PEMEX, Petroecuador) set within their business policy objectives to reduce greenhouse gas emissions within the oil and gas value chain, setting emission reduction targets by operational area: production, transport and refining.

According to the information provided by the aforementioned companies, the main source of emission is combustion, followed by indirect emissions corresponding to electricity and steam, followed by burning teas and finally emissions from vents in processes, which refer to the controlled release of gas in the production wells to facilitate the process of crude oil extraction or pressure release according to operating conditions.

*Process optimization: involves*the reduction ofveins; methane leaks, sulfur hydrocarbon compounds; reduction and optimization of burning
Low carbon energy diversification: energy diversityis understood astheuse of different sources ofenergy(renewableandnon-renewable), to meetthe needs. By replacingenergy sources withhigh carbon footprints such as coal, crude and diesel with low-carbon sources such as natural gas, biofuels or electricity from the national electricity grid and increasing the participation of such sources, reducingemissionsspecifically; adems, promoting energysafety through flexibilityand increasing the energyscienceofprocesses.

Carbon capture, use and storage: an alternative to mitigate large volumes of CO2. By making use of the CO2 captured in injection processes for improved recovery, the resulting mitigation process is cost effective, obtaining an economic benefit from commercializing the CO2, and obtaining a greater number of barrels of crude from mature wells that have declined in their production of oil and thus avoid CO2 in the environment.

Oil activities must be carried out under a criterion of prevention and responsible management of all the environmental and social impacts it generates. First, it ensures compliance with the social legal obligations committed to the environmental authority, and secondly, other uneseen impacts identified in the development of activities are managed, allowing the community to feel calmer and safer.

There are usually phenomena of deterioration of ecosystems and the social fabric, by deforestation and pollution, by discouraging the productive vocation of the region and petrolization of the regional economy, by changing the dynamics of employment, by increasing conflictivity and by alteration or extinction of cultures, among others. In addition, in recent years, the perception of many people towards the oil industry has become increasingly negative, preventing oil and gas search, extraction and production activities, in order to maintain the existing socio-environmental balance and not to take the risks involved in these operations.

In the world, fracking operations have been increasingly rejected, with harmful effects on underground aquifers, increased seismic activity, environmental pollution, among others, generating citizen passions and debates at all levels (Barrio & Pérez, 2009; Non-hydraulic fracture, 2014). Therefore, in mandatory to mitigate the environmental and social impact, with innovation and technology, generating efficient and measurable programs to manage impacts and achieve a new balance in communities, where it is also part of the challenges for the oil and gas industry in association with the academy.

In terms of human talent, we have great potential, but it requires a great deal of effort to articulate and respond to the needs of industry and communities. In the global oil and gas sector, there has been a phenomenon of generational change – Big Crew Change (Loh, 2013), in which a high percentage of professionals over 50 years of age and with great experience (34%) has begun to retire, followed by a small percentage of human talent between 36 and 50 years (8%), which with less expertise has had to address highly complex projects, and which in many cases represent delays, increased risks, higher project costs, among others (Schlumberger Business Consulting – SBC, 2012; Dupre, 2013).

This situation originated in the mid-1980s, at a time of low crude oil prices when the oil and gas industry was not labor competitive and as a result student in oil engineering, geology and other careersdeclined. Therefore, anotherof the challenges that the academy has is to offer aquantity of graduates in geosciences and engineeringofpetrorleos that through anaccelerated learning curve, achieve the technical and humanskills required by theindustry andcan beincorporated and physically by the sector,which for its part has had to designnon-traditionalretentionand recruitmentstrategiesto ensure the recruitment of the scarcespecializedhuman talent, with the least possible impact onthe normal management of its operations (Barna, 2010).

In the same sense, companies must be increasingly demanding in topics of ethics and accepted behaviors, which must also be reinforced from the academy. While it is true that it is necessary to increase reserves, increase the recovery factor, optimize costs, evacuate heavy crude oils, improve fuel quality, efficiently manage projects, incorporate and develop state-of-the-art technologies, achieve process safety, care for the environment, renew generational tables, among others, in a sector where the objective is to maximize economic value, challenges related to behavior, principles and values, can become the most important.

A new environmental planning model

5 SCIENTIFIC RESEARCH FOR THE IMPULSE OF HUMAN DEVELOPMENT

The word research comes from the Latin voices in, *vestigium, ire,* which means going after the track, and which can be explained as a way to show reality to indecent, question or interpret it. In practice, research is conceived as a way to know reality through a thoughtful, systematic, controlled and critical method or procedure, which allows the interpretation of facts and phenomena, the establishment of relationships, the application of laws, the approach of problems, the search for solutions, and the creation of conditions for changes and transformations

The processes and role of research begin from the very beginning of human life and develop further in the following stages. However, over time the social environment produces or generates guidelines and typecasts of defined defeaters that are imposed on us without allowing the curiosity, creativity, observation and exploitation of themselves that human beings by nature possess, that is, the innate gift of research.

This is where globalization, in its good sense and proper application, plays an important role in the development of Latin American societies in economics, culture, science, academic level, among others. It is clear that development not only goes to production growth as a form of economic income, but must be represented in the intangibility of societies. That knowledge will remain an undeniable heritage in our generations.

For this reason, technicians and administrators are required to live and apply moral values and principles, respectful in their personal and corporate relationships, responsible in their daily and work life, transparent and whole in their thinking and acting. Human talent must be world-class, enthusiastic, innovatively thought-out, acting in advance to protect one's integrity, that of others and that of the environment, with propositional judgment, discipline and the ability to share knowledge and information to learn and grow professionally and personally. The success of the oil sector in the economic, social and environmental dimensions requires a collaborative environment, with competent, ethical and entrepreneurial leaders and technicians, to ensure the creation of value and individual prosperity and all stakeholders (Porter & Kramer, 2011).

The relationship between science, technology and development and innovation is always complex and even more so when analyzed around the conditions of developing countries, where science and technology do not take on the deserved importance in terms of economic support, demonstrating a huge dichotomy among private industry with government institutions.

As South America is a region dependent on energy and mineral resources, we make little effort in finding the solution in savings and innovation that places solutions to the different industries that need our raw material on the market, and that are directed at accelerated growth in their (GDP) Gross Domestic Product (Graph 2), leaving us behind in regional growth on two clear aspects :

- Generation of researchers and their incentives,
- Lack of budget and/or interest in those within public or private educational institutions and bodies.

Where regional interconnectivity is vitally important in creating a knowledge-based society, and a growth in human development.

Figure 2. State commitment to research and development in science and technology

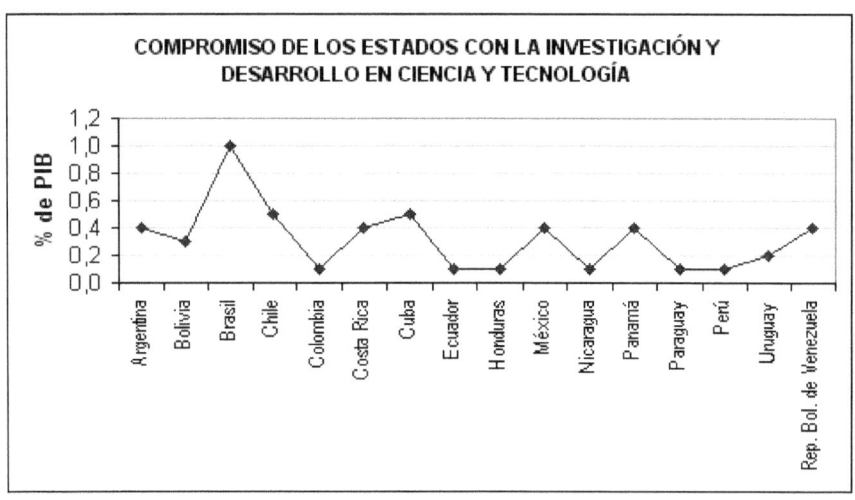

Source: Extracted *from Science and Technology in Latin America* Electronic Edition (2007).

The idea of associating green technology with science, technology and innovation is so that, from thought, design, production that generate solutions and benefits without unpleasant consequences to the planet and of course to the nature and health of society at large is made with tangible and long-lived savings.

These savings can achieve reduced maintenance, prevent failures, increase product life, lower power consumption, and improved efficiency. The key issue would be in the realization of these benefits, not only in basic research, but in the transfer of knowledge from the designer to the user's community, respecting intellectual law.

LATIN AMERICAN ENERGY SAVING

Unfortunately, in Latin America, they have not spent enough time, due to various factors, or the attention that deserves the development of this technology, except for some specific cases. However, South America should be the region that invests the most in research, and development of this technology, as it is an oil and mining region par excellence, this condition should be taken as an incentive to create new energy and environmental solutions on a global scale.

In the South American approach, it should be preferred to invest in disciplines that specifically influence tomorrow's quality of life and development aspirations, taking into account the well-being of society over monetary benefit, which will undoubtedly exist.

This implies initiating an expenditure plan from the present, gradually and increasingly addressing the meeting of the goals, framed in a regional Science and Technology Plan for Development of the first three decades of the century, in the case of States, and a Common Goals plan in Science and Technology for Latin America. In the latter case, efforts must be exhausted in integration and cooperation to articulate Community bodies, in the same time horizon. To this end, the commitment in research to GDP in the region must be increased, as these targets are not balanced today, as shown in Graph 2.

There is a well-marked bias with Brazil, Mexico, Costa Rica, and Chile at the tip, with respect to the countries that make up the southern hemisphere, according to the Global Innovation Index 2019, developed by the World Intellectual Property Organization (WIPO) these were the latin American nations best positioned

In this regard, some countries have understood that energy abuse can be avoided, making adequate and conscious use of the existing technology today, giving importance to continuous innovation to that place, with a credit to the lack of fear of making mistakes in the innovation process, giving them opportunities for improvement based on the results shown, eliminating punishments for mistakes, following the philosophy of Silicon Valley thinkers, being aware that we have ingenuity and creativity, only the socio-political-economic commitment to their exploitation is lacking.

The figures are conclusive, countries with high levels of development also record high levels of spending on research and technological development as shown in Figures 2 and 3. The challenge, then, relates to finding possible and realistic ways to improve the situation in our countries.

Figure 3. Thirty first countries in scientific production. 2018

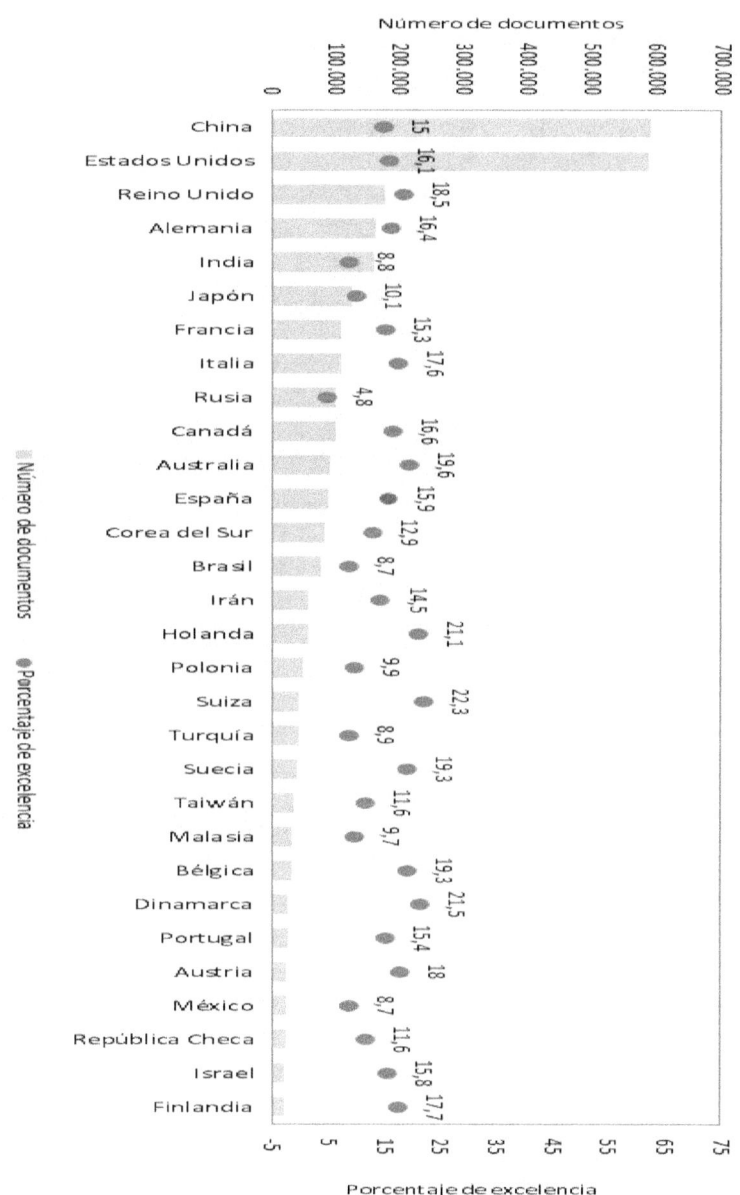

Source: INE, Eurostat and OECD

Figure 4. R&D spending as a percentage of GDP in the world. 2018

A new environmental planning model

Number of documents: Sum of scientific articles, minutes of congresses and annual reviews Rate of excellence: Indicates what percentage of scientific publications in a country or institution are included in all 10% of the most cited articles in its area. (Text extracted from FECYT Spanish Foundation for Science and Technology)

Source: SciVal-Scopus.

A new environmental planning model

6 A NEW PLANNING MODEL

It is clear that societies need an urban - industrial environment for their existence, but never should our very societies compromise our planet. This is where the use of innovation and technology to generate great energy and therefore economic savings is essential, with a balanced environmental model that translates into social benefits.

For this achievement, technically function planning (with positivist and rationalistic bias) is required, it must have a socio-ecological approach to bridge the gap between rapid technological and economic development, which must bet on the decrease in social entropy of the system itself. This establishes the need for a new type of planning: Sustainable or environmental according to some, ecological, spatial, strategic, among others.

The new planning model should seek to bring knowledge closer to action, i.e. without forgetting the future, to emphasize current processes. This new model must then be "normative, innovative, political, negotiating and based on social learning". (Friedman, J. 1992).
This environmental planning can be conceived as: "The instrument aimed at planning and programming the use of the territory, productive activities, the organization of human settlements and the development of society, in consistency with the natural potential of land, the sustainable use of natural and human resources and the protection and quality of the environment". (Salinas, E. 1991, 1994, 1997 and 2005).

Environmental planning seeks to organize socio-economic activities in space, respecting their ecological functions in a way that promotes environmental sustainability and sustainable development (Instituto Brasileiro do meio ambiente e dos resources naturais renováveis - Ibama, 1995).

This systemic conception of environmental planning proposes that there cannot be a long-term ecological balance along with critical socio-economic situations such as: poverty, malnutrition, illiteracy, among other social ills; just as socio-economic development is not possible without it being in line with the availability and renewal of natural resources on the one hand (so-called natural capital by some authors) and the development of productive forces on the other.

On the other hand, the incorporation of sustainability into the productive and social process depends on us achieving energy efficiency in the landscape environment, using more appropriate technologies, achieving social equity, adjusting growth to available natural potentials and resources, and adapting and accountability in decision-making. In addition, we must strike a balance in the intrinsic characteristics of the landscape as a Geo-ecological and socio-cultural support of sustainability. This will enable the conception of sustainable landscape seen as "a place where human communities, resource use and load capacity can be maintained in perpetuity" (Matthew J. 1997).

The challenge is for future generations, who, from Research and Development and Technology and Innovation, look for meeting points for us to maintain socio-environmental balance, where the parties (Societies, Ecosystems, planet, living beings among others) maintain synergy for their existence and longevity.

STRATEGIC ENVIRONMENTAL PLANNING

The need to incorporate the environmental dimension, as well as citizen participation in the instruments of territorial order and mainly those of urban areas, requires the use of new planning methodologies that promote the use of the territory in a sustainable way, that is, reconciling economic, social, political and environmental interests both the short and long term. In the field of environmental assessments, the methodological instrument called Strategic Environmental Assessment is increasingly being implemented in order to complement environmental protection and quality with economic and social development. This approach makes it easy to have a broad-spectrum vision that allows you to quickly assess the impacts of territorial planning within the overall framework of proposed environmental quality objectives.

Environmental planning is configured as a synchronous and organized decision-making process in a delimited geographic space, which possesses and processes a specific and significant knowledge of the territory in environmental assets, which are real, dynamic and changing and that, ordered and organized, converge in a systemic direction to the integral vision of the planned object. However, the time of planning, and especially of the territorial environmental dimension, is when information on the environmental empirical knowledge of the territory needs to be transformed, that is why We must incorporate Elements framed in a baseline, which interacts with three fundamental elements:

Physical environment: Geology - Geomorphology - Hydrogeology - Hydrology - Edafology - Climate - areas of natural risks - Historical and cultural Middle Landscape - National Monuments - Architectural Heritage - Archaeological Sites - Historical Heritage.

Environment: Biotic - Flora and vegetation - Fauna of terrestrial and aquatic vertebrates – Limnology.

Social Environment: Populated centers - Demographics - Economic activities in the Middle Built - Land use - Equipment - Infrastructure and services - Road infrastructure.

All these variables are mandatory for the ordering of a territory, the inclusion of the environmental variable represents guaranteeing over time the quantity and quality of renewable and non-renewable natural resources, as well as the environmental services available therein and therefore, having an environmental policy is established as a vector of environmental sustainability of the territory and within it, environmental planning contributes to the very strategies of the ordering of a desired system, logical and flexible, that is, an instrument to guide actions and criteria in the management or sustainable use of the territory and the construction of spaces, subjects and territories simultaneously (Vega L. , 2002; Wernes, 1995). In this sense, planning in the order of the environmental dimension, proposes a qualitative and quantitative knowledge of the very composition of the ecosystem and a rationality, as well as the efficient use of resources, in terms of the potentialities, limitations and characteristic of the environment as the basis for the functioning of the natural system (Wernes, 1995), which can collectively make decisions of binding actors on the environment so that unjustifiable damage does not occur and there is a global sustainable development of the territory, within a reference framework that establishes specific guidelines and measures of intervention (Leitmann J. , 1999; Millar D. , 2005; Sheila S. , 2004; Rivas, 2002).

T&I - R&D & ENVIRONMENT

Today it is impossible to think of a world without information and communications technologies (ICT). Its increasingly widespread use has changed the lives of many people and boosted economic growth, but its contribution to greenhouse gas (GHG) emissions continues to grow. However, the use of ICTs provides great opportunities to reduce these emissions, especially in industries such as power generation, waste disposal, construction and transport. (Malcolm Johnson Director of the Telecommunication Standardization Office of the International Telecommunication Union (ITU) 2011).

Concern for the environment is not a passing fad or typical only of environmentalists. Today, the care of the planet has become a matter of particular relevance to both citizens, civil organizations and governments. Where Latin America and the Caribbean are not left behind in this global trend, and begins to incorporate tools to combat climate change.

In the context of current globalization, no industrial, commercial or service policy, just as social policy will succeed if it does not know the need to incorporate the principles of sustainable development as guides to economic growth. Gone are the times when natural resources were exploited and it was produced to the fullest without considering the environmental impact that was generated. At this time, it is necessary to adopt appropriate environmental management methods in response to drastic changes in industry production systems; marketing channels for products and service distribution networks, as well as the impact that any technological insertion would have on the social collective within this century and in the coming century.

T&I – D&I are essential to help our countries adapt and prepare for climate change, action needs to be taken to mitigate its effects and plan for the future. In addition to providing education and information through transmissions, the Internet and other means, it is worth mentioning the importance of remote monitoring of the Earth by satellite and sensors in the soil and seas. This can be used, for example, to extract data on deforestation or crop patterns that indicate possible food shortages. In addition, ICTs are vital when it comes to warning about natural disasters that may arise as a result of climate change, as well as to address its effects, by allowing humanitarian teams to respond in different ways.

The importance of climate change in the region, as well as the search for solutions that can minimize environmental impact through T&I – D&I is highly relevant, as Latin America and the Caribbean face a constant danger of events such as floods, hurricanes or droughts as a result of climate change. Recent studies indicate that T&I – D&I can help reduce global greenhouse gas emissions by approximately 15% since 2020, through initiatives such as video conferences, e-commerce, e-government or smart buildings, which arose from the need to keep our productions moving, in a way the pandemic has collaborated with the planet , with great respect to the relatives of the deceased, this sacrifice must not go unnoticed.

Another relevant initiative to reduce the impact of pollution is the proper management of electronic waste. The rapid emergence of new technologies generates a high number of e-waste, which can be recycled and reused, either in whole or in part, which gives relief to the planet's over-exploited deposits and maintains costs in the demand-economy ratio.

LATIN AMERICA AND THE CARIBBEAN AND ENVIRONMENTAL TECHNOLOGIES

In Latin America, companies that have failed to integrate environmental criteria with the competitive strategy still prevail, and the dichotomy between public policies and private strategies prevails, which shows a delay in environmental technology, unlike European and North American companies that have exceeded the threshold between the production strategy and environmental actions, as a mechanism for gaining competitive advantage.

However, it is worth mentioning that the successful development and operation of companies, require continuous assessments of opportunities, risks and trends. These assessments were previously conceived under economic, political and social criteria, but today the environment is emphasized as a successful agent, where European and American union companies can put sustained pressure on the closure of future businesses.

Not everything is lost, there are action plans in the region where it is necessary to sustain and harden, from the point of view of sustainability, the approach so that, through comprehensive management policies of electrical and electronic waste, based on the positive relationship of the actors, developing mechanisms for coordination between the different sectors: public, private, decentralized and civil society. In addition, the scientific and operational use of T&I – D&I makes it possible to scientifically understand and detect natural phenomena that generate natural risks and disasters. For this reason, these technologies are used to take preventive and reactive measures, and establish early warning systems.

Latin America and the Caribbean are already implementing a number of initiatives aimed at promoting integration and minimizing the impact of climate change through T&I – D&I, as well as others – most – that focus on e-waste management and recycling. Below are some of the actions carried out by the countries of the region in this area.

Argentina

In response to the problem of technological waste, the "Seminar on Sustainable Waste Management of Electrical and Electronic Equipment *Waste*" has been held since 2008, with the aim of addressing the problem and promoting a waste management program for electrical appliances that promotes the collection, selection, disposal and valuation of parts and materials susceptible to reuse and recycling in new industrial processes, or their donation. In addition, "National Parks and Interactive Schools" has been developed, a program of computer equipment, satellite internet connection and face-to-face and virtual training, with the aim of reducing the digital divide in the communities involved, while promoting the conservation of water, fauna and flora and promoting sustainable development through education. The program promotes environmental education through ICT, and has also placed schools, by connecting them to the Internet, as the epicenter of various activities, both educational, social, cultural and recreation.

Brazil

In May 2010, Brazil's Ministry of the Environment and Cempre NGO signed an agreement for the creation of the country's first inventory of electronic waste production, collection and recycling. The objective of the agreement is to measure the generation and destination of electronic waste in Brazil, as well as to assist in the generation of public policies and identify the main bottlenecks in the recycling chain. In the country there are also electronic recycling initiatives, such as the one carried out by the University of Sao Paulo, where in 2009 an electronic waste recovery and processing center was opened. On the other hand, the CI project, of computers for inclusion, has been operating since 2004 with a recycling network of discarded ICT equipment, which are refurbished and then donated to telecenters, schools and bookstores in the country. In addition, since 2004 the International Fair of Environmental Technology (FIEMA), under the auspices – since 2007 – of the PROAMB Foundation, an organization with 20 years of experience in the environmental area, has been held since 2004. This fair seeks to carry a growing number of companies and organizations, national and international oriented to technological production, solutions and services focused on the environment and sustainable development. In the different segments participating in Fiema Brasil there are exhibitors working on introducing ICTs as a solution to solve environmental problems, as well as those for the disposal and recycling of computer equipment.

Plurinational State of Bolivia

In the Plurinational State of Bolivia, the Environmental Information System (SIA), belonging to the National Chamber of Industries, has centralized and computerized Bolivian environmental information in a single system, with alphanumeric and cartographic components. In addition, the new Telecommunications Law, which is based on five axes, one of which is the environment, is being developed within the Vice Ministry of Telecommunications; for this reason, attention is paid to issues involving the correct disposal of electronic and telecommunications equipment, electromagnetic emissions, the deployment of radio bases or other communication bases within protected areas, as well as the development of environmental data sheets for each civil construction corresponding to telecommunications, among others. There have also been studies in the country by NGOs, such as the Swiss Contact Foundation or the Quipus Foundation, that diagnose the potential environmental impact of both existing electronics in Bolivia and the waste they generate; In addition, the REDES Foundation represented Bolivia in the e-waste working group in the regional action plan for the eLAC2010 information society, until the establishment of the new eLAC2015. On the other hand, some municipalities have carried out electronic garbage collection initiatives since the mid-2000s.

Chile

Chilenter is a Chilean foundation, whose motto is to contribute to the social use of technology and that is an environmentally sustainable manager, since it incorporates in its work the main guidelines recommended at the international and national level for the management of electronic waste. In the country, Chilenter is the main player in the field of the reuse of obsolete technology, with the capacity to refurbish approximately 15 thousand computers per year. The refurbishment process is to enable de-de-de-source equipment through comprehensive technical and administrative procedures, including diagnosis, part and part selection, computer assembly, operating system installation and configuration, and equipment quality control. For its part, the Committee for the Democratization of Computer Science in Chile, CDI, through its *campaign "Donate your computer",* collects equipment that is no longer in use to be refurbished and installed in schools and telecenters. In addition, in Chile there is the web portal SINIA – National Environmental Information System," administered by the Ministry of the Environment and consisting of a set of databases, equipment, programs and procedures dedicated to managing information about the country's environment and natural resources, in an integrated and interpretable way. Through this portal you can directly access the different information systems that are currently integrated into the SINIA.

Colombia

In 2013, there will be between 80 and 140 thousand tons of electronic waste in Colombia corresponding to disused computers, according to mmsI. That is why the National Centre for Electronic Waste (CENARE) in that country works to achieve the reduction of these figures while promoting ICT in the classroom. Thus, through donations, the center has received 211 thousand computers, of which 130 thousand were donated to schools and the rest became waste. With these, *CENARE* also works on the robotics and automatic educational project, which seeks to integrate children from public schools into science and technology by building robots with disused elements of unarmed computers. Colombia's technological waste management programme was highlighted by UNESCO, which in a report cited the country as an example of good practice in this area. In addition, since 2001, a tax exemption has been applied in the country to encourage the incorporation of technologies that benefit the environment and health, and electronic waste collection campaigns, particularly mobile phones and computers, are regularly carried out.

Costa Rica

In response to ICT's action on the environment, the construction of the "IT Friendship Meter with the Environment" was initiated in Costa Rica in 2009, developed by the Scotiabank Information Systems Center together with the Costa Rican Technology Research Club. Its objective is to measure the impact of ICTs on the environment and promote comparable information between different organizations, in order to generate a change in behavior and reduce that impact. The development of this meter is based on the idea that it is essential that the ICTs used are environmentally friendly, to improve the efficiency of organizations and the quality of life of all. In addition, there is the organization Costa Rica neutral, which on its website allows through a simple virtual calculator estimate the amount of emissions of a house, office or store. On the other hand, the country's National Emergency Commission has a communication system to minimize the impact of natural disasters through early warning, which, through radio, Internet and satellite systems, keeps the community and the CNE alert to possible natural threats. In addition, the Costa Rican Vulcanological and Seismological Observatory uses text messages to keep the community informed.

Ecuador

According to MMSI information, there have been multiple private initiatives in Ecuador, especially from mobile phone companies, seeking to recycle electronic devices. In addition, the Superintendency of Telecommunications *(SUPERTEL)* has recommended establishing regulations for smart appliances in terms of energy conversion, type of plugs used and reuse of devices; ON the other hand, SUPERTEL seeks to promote the integration of technologies for the provision of services and the development of recycling and the safe disposal of technological waste. Since the end of 2009, a comprehensive management and recycling initiative for electrical and electronic waste has been developed – through the company Vertmonde. During 2011 this initiative will be re-developed in the city of Quito and Guayaquil in the first semester. In addition, a recycling campaign will be launched with the participation of the entire technology equipment marketing community, in which the waste generated by wholesalers and its distribution channel will be collected. By the end of 2011, it is expected to collect more than 90% of the waste generated or collected by this community, as well as to have a baseline of the amount and type of waste generated by this sector, in order to implement the same model at the national level.

Cuba

Information is essential in the fight against climate change and the need to care for the environment. That is why, in Cuba, it has been centralized in websites aimed at delivering environmental data, which offer environmental statistics, publications on the subject, links to related sites, indicators of electricity consumption in ministries and information on projects, among other things. One of them, the Environmental Education Portal of Cuba, is supported by UNESCO's regional office in that country, and seeks to achieve the integration of results, promote greater dissemination of results and continue to increase and share successful experiences in the environmental field.

Peru

According to MMSI, there are three formal companies in Lima that collect e-waste; however, they only process 3% of the 15,000 tons of cell phones and computers whose lifespan ends each year in Peru. For the same reason, the Ministry of the Environment has decided to support private campaigns - a *joint project of IPES,* the Ministry of the Environment and the municipality

– and since June 2010 a pilot program has been carried out in the municipality of Santiago de Surco, focused on supporting electronic garbage collection campaigns, which is to be replicated throughout the country. Peru also has the *SINIA*, National Environmental Information System, a network that facilitates the systematization, access and distribution of environmental information, as well as the use and exchange of it. Through its website the population can access information composed of environmental indicators, thematic maps, documents, reports on the state of the environment and environmental legislation.

Uruguay

The implementation of **the CEIBAL** plan in Uruguay – the experience of one computer per student – has paid excellent dividends in education, but presents a challenge for the care of the environment. This is why various e-waste recycling initiatives are carried out in the country. One of them is the one developed by the logistics department of the Ceibal Plan, which works with a logistics services company to address this problem. Among other things, the department analyzes the amount of electronic garbage that is and will be generated by the Ceibal Plan, with the intention of reusing the parts that can be recycled from the laptops delivered to the children. In this way we also want to minimize future purchases of specific parts of computers that are required to repair *so-called ceibalites*. There are also companies such as Crecoel – Cooperative for the Recycling of Electronic Components –, an undertaking that seeks the dismantling and recovery of materials from electronic equipment and components. The cooperative charges companies and public entities for this service, but not those with less than one cubic meter of waste.

Bolivarian Republic of Venezuela

Since 2007, the Venezuelan government has implemented a social and economic development plan, focused on deepening specific public policies. These initiatives include the redesign of the national science, technology and innovation system to support programmes that use ICTs for the environment, as well as those that help education in this area. In addition, Venezuela seeks to establish national alert systems that use ICT as a warning tool, as well as automated climate stations that promote the exchange of critical information. One problem facing Venezuela, the result of climate change, is the meltdown of the glaciers of the Sierra Nevada in Merida. To monitor this phenomenon there is the Merida Bioclimatic

Network, which uses a web-based bioclimatic information system, which allows easy access to raw data from each participating station of the network, as well as the possibility to send data from both conventional and automated stations to a central collection site, using a web interface. In addition, it provides a tool for querying climate data by station, geographic location, time period, among other variables, and access to processed data, such as charts, maps, tables, and animations

The Caribbean

An excellent regional initiative involving Caribbean countries is *the Caribbean Information Platform on Renewable Energy (CIPORE)*, an information and communication system on the regional use of renewable energy, which aims to gather all of each country's information regarding renewable energy at a single point of access. The website http://cipore.org has multiple information on the use of renewable energies, which can also be filtered by solar, geothermal, wind, nuclear, water and biomass type. On the page you can find links to the agencies, ministries of energy and universities of each country concerned about renewable energies, as well as find detailed information on different renewable energy initiatives and projects in the Caribbean. The page features information in English, French, Spanish and Dutch.

PENDING POINTS TO INTEGRATE AND INCLUDE

The Science and Technology strategy has been considered fundamentally and important to lay the foundations for a new articulation between all sectors. Therefore, scientific and technological development should be oriented to improve the existing socio-economic situation, using the human potential and natural resources that are possessed with a long-term and integral vision. In this process, it is important to intervene by the State, in terms of the management of transparent and well-explicit policies that control and regulate the implementation of the negotiations that have been materializing through the Free Trade Agreements and everything that concerns environmental ordinances, in the short, medium and long term.

The management of these practices involves the strategic planning of companies and the definition of their needs, integrating the education sector into the analysis of R&D as a generator of innovations in the market, from which knowledge is deduced, access to technological tools, among others, in order to obtain a dynamic balance between the demands of society and the availability of such environmental goods.

In October 2010, the ITU Plenipotential Conference adopted a new resolution for the role of T&I – D&I and environmental protection, which identifies the need to assist developing countries in leveraging these technologies in favor of combating climate change. Subsequently, at the Cairo Symposium, and on the basis of the discussions held there, the roadmap was created with the following recommendations for the use of technologies for the environment.

Step 1: Share best practices and increase sensitivity about the benefits associated with using green technologies.
This step seeks to stimulate and, where possible, stipulate that there be a broad exchange of best practices and information to maximize the dissemination of green technologies and intelligent technological solutions in the public and private sectors. It also seeks to promote teaching on green technologies and to increase awareness of environmental implications.

Step 2: Demonstrate success and viability.
The development of methodologies and indicators to measure and monitor environmental impacts on the life cycle of services and technological devices, including measurements relating to greenhouse gas emissions, is to be encouraged. In addition, this step aims to use pilot and flagship projects to help spread the most promising smart solutions in sectors such as buildings, transportation and energy.

Step 3: Involve the private sector, civil society and the academic community.
The paper proposes that these sectors play a leading role in protecting the environment through innovation and the correct use of green technologies to address climate change. Therefore, it seeks among other things that promotes environmentally friendly and socially responsible research and development (R&D).

Step 4: Promote national, regional and international cooperation.
Cooperating at these levels is essential to foster a path to low-carbon sustainable economies, this step arises. In addition, it would enable greater green investment and sustainable management of natural resources, as well as the development and dissemination of clean technologies. It also seeks to encourage developed countries to assist developing nations in their efforts to include and adopt policy reforms towards greener growth.

Step 5: Integrate government policies, climate change, environment and energy.

This step raises the need to bridge the gap between technological development, the environment and energy experts, as well as policy makers, to enable the integration of environmental and energy policies. On the other hand, it seeks to integrate the use of technologies into the adaptation of national plans to make use of them as a tool to address the effects of climate change and minimize the environmental impact of public administration through policies, applications and services. Finally, it proposes the establishment of transparent policy objectives to improve government strategies, with monitoring and evaluation of compliance.

Step 6: Develop and implement a national strategy of technology and green innovation for growth.
It argues that such a strategy should be had at the national, municipal and community levels, as well as individual organizations. The green strategy needs to be seen as a component of the national development strategy, and the use of technologies in support of environmental management must pass through all sectors of the economy and levels of society. Technical support should be provided to countries that require it, especially those in development, to help them formulate and implement green strategies.

A new environmental planning model

CONCLUSION

We are at the best time to live up to countries with constant development, both in human resources with unrestricted support for imagination, including failures that are intrinsic to research, with competent and constant support.

Scientific work allows us to establish the understanding and explanation of universal causes, principles, processes and laws, in order to increase the relationship between man and nature, regardless of the surrounding political and social context, thus achieving the satisfactors of needs common to most human beings. The scientist creates nothing in the absolute sense, since it is the Creator of the Universe who placed man in a world full of wonders that simply had to be discovered and developed to solve the problems that have gradually appeared through human history.

On the other hand, technology is about applying scientific and empirical knowledge to solve the current problems that are defined according to the economic, political or social needs of a particular society or group. Therefore, we can say that a country's technological development does not involve using the technologies of developed countries but trying to meet its needs with its own human and material resources.

For this we need to conclude public authorities and private entities with a joint effort to seek the closure of gaps, with a legislative search where we have common benefits, involving increased gdp in R&D and R&T, construction of laboratories in our universities (public and private), business policies with futuristic projection, thinking of an attractive patent for use on all continents.

The payment to this collective effort will be to guarantee future generations the power to enjoy the benefits of developing technologies and applying innovations for improvement in all aspects of everyday life.

BIBLIOGRAPHIES

Bustillo Revuelta, M.; López Jimeno, C. (1996). *Recursos Minerales. Tipología, prospección, evaluación, explotación, mineralurgia, impacto ambiental.* Entorno Gráfico S.L. (Madrid).

CEPAL, *TIC y medio ambiente.* Newsletter. Marzo 2011

Friedmann, Jhon (1992) *"Empowerment: The Politics of Alternative Development"* Primera edicion (1990). Editora Blackwell Publishers. Dublin, Irlanda.

Girón, Alicia (coord.) (2014). *"Trilogía: Cómo sembrar el desarrollo en América latina, Colección de Libros Problemas del Desarrollo".* Instituto de Investigaciones Económicas-UNAM México.

Isabelle Barois, Silvia M. Contreras Ramos, Benito Hernández Castellanos, Martín de los Santos, Froylán Martínez y David R. García. *El suelo y el petróleo.* (2018). Instituto de Ecología A.C. D.F México

Miranda Vidal, Julio: (2007). *"Ciencia y tecnología en América"* Latina. Edición electrónica.

Moustafa Gadalla (2007) *La Cultura Revelada Del Antiguo Egipto.* Tehuti research Foundation. EEUU

Organización Mundial de la Propiedad Intelectual (OMPI): (2019) *Innovar para un futuro Verde.* Extraído de https://www.wipo.int/ip-outreach/es/ipday/. Suiza

Peñaloza Acosta, Mónica, Arévalo Cohén, Freddy, Daza Suárez Roberto; *Impacto de la gestión tecnológica en el medio ambiente.* Revista de Ciencias Sociales v.15 n.2 Maracaibo jun. 2009.

Salinas Chávez, Eduardo (2005). *"El Desarrollo Sustentable Desde la Ecología del Paisaje".* Extraído de http://www.gobernabilidad.cl/modules.php?name=News&file=article&sid=796.18 páginas. DF México.

Susana Chow Pangtay (1987) *Petroquímica y Sociedad. Fondo de cultura económica*, S. A. de C. V. D.F México.

WEB REFERENCES

https://www.english-heritage.org.uk/

https://ejatlas.org/

A new environmental planning model

A new environmental planning model

ABOUT THE AUTHOR

Alexis José López Delgado (1978). He developed a particular interest in geo-scientific research influenced by presocratic philosophy. He managed to harmonize science and spirituality, achieved personal transformation after an intense self-assessment. He understood that the science-religion relationship is necessary to ensure the longevity of humanity with an implicit seal of universal common good, never to satisfy personal benefits, rather to meet the needs of majorities.

www.ingramcontent.com/pod-product-compliance
Lightning Source LLC
Chambersburg PA
CBHW070434220526
45466CB00004B/1673